Rapid Assessment Program

A Rapid Biological Assessment of the Kwamalasamutu region, Southwestern Suriname

Edited by
Brian J. O'Shea, Leeanne E. Alonso, and
Trond H. Larsen

T0086660

RAP
Bulletin
of Biological
Assessment

63

CONSERVATION INTERNATIONAL - SURINAME

CONSERVATION INTERNATIONAL

LBB/NB NATURE CONSERVATION DIVISION, SURINAME

SBB SURINAME FOREST SERVICE

STICHTING MEU

PANTHERA

STINASU

AMAZON CONSERVATION TEAM (ACT)

ALCOA FOUNDATION

The *RAP Bulletin of Biological Assessment* is published by:

Conservation International
2011 Crystal Drive, Suite 500
Arlington, VA USA 22202

Tel : +1 703-341-2400
www.conservation.org

Editors: Brian O'Shea, Leeanne E. Alonso, and Trond H. Larsen
Maps: Krisna Gajapersad
Design: Kim Meek
Cover photos:
[top] A stream runs through the forest at the Sipaliwini RAP survey site in the Kwamalasamutu region.
(T. Larsen)

[lower left] The Giant leaf frog (*Phyllomedusa bicolor*) is the largest tree frog in the world. Its skin produces a
highly toxic secretion, which protects this slow-moving amphibian from predators. (P. Naskrecki)

[lower right] The cichlid (*Apistogramma steindachneri*) is a colorful fish, commonly kept as an aquarium pet.
Both males and females protect the developing young. (P. Naskrecki)

ISBN: 978-1-934151-50-1
©2011 Conservation International
All rights reserved.

 Printed on recycled paper.

RAP Bulletin of Biological Assessment was formerly RAP Working Papers. Numbers 1–13 of this series were
published under the previous series title.

Suggested citation:
O'Shea, B.J., L.E. Alonso, & T.H. Larsen, (eds.). 2011. A Rapid Biological Assessment of the Kwamalasamutu
region, Southwestern Suriname. RAP Bulletin of Biological Assessment 63. Conservation International,
Arlington, VA.

This RAP survey and publication of this RAP report were made possible by generous financial support from
the Alcoa Foundation.

Table of Contents

Participants and Authors

Leeanne E. Alonso, Ph.D. (ants)
Rapid Assessment Program
Conservation International
2011 Crystal Drive, Suite 500
Arlington VA 22202 USA
leeannealonso@yahoo.com

Aritakosé Asheja (park ranger)
Amazon Conservation Team
Kwamalasamutu Suriname
info@actsuriname.org

Olaf Bánki, Ph.D. (plants)
Netherlands Centre for Biodiversity Naturalis
National Herbarium of the Netherlands
P.O. Box 9514, 2300 RA Leiden, The Netherlands
olaf@banki.nl

Angelica Benitez (contributing author, large mammals)
Panthera
Calle 93Bis, #19-40, Oficina 206
Bogotá Colombia

Chequita Bhikhi, M.Sc. (plants)
Netherlands Centre for Biodiversity Naturalis
National Herbarium of the Netherlands
P.O. Box 9514, 2300 RA Leiden, The Netherlands
cheqbr@gmail.com

Natalia von Ellenrieder, Ph.D. (dragonflies and damselflies)
Plant Pest Diagnostics Branch
California Department of Food and Agriculture
3294 Meadowview Rd.
Sacramento CA 95832 USA
natalia.ellenrieder@gmail.com

Klassie Etienne Foon (tree spotter)
SBB/Suriname Forest Service
Martin Luther Kingweg 283
Paramaribo Suriname

Donovan Foort (game warden)
LBB/NB Nature Conservation Division
Cornelis Jongbawstraat 12
Paramaribo Suriname
the_scorpion@live.nl

Krisna Gajapersad (large mammals, local logistics)
Conservation International – Suriname
Kromme Elleboogstraat 20
Paramaribo Suriname
k.gajapersad@conservation.org

Mercedes Hardjoprajitno (water quality)
Anton de Kom University of Suriname
Leysweg 9
Paramaribo Suriname
mercedes.hardjoprajitno@gmail.com

Rawien Jairam (reptiles and amphibians)
National Zoological Collection of Suriname
Anton de Kom University of Suriname
Leysweg 9
Paramaribo Suriname
rawien_2000@yahoo.com

Reshma Jankipersad (plants)
SBB/Suriname Forest Service
Martin Luther Kingweg 283
Paramaribo Suriname
rwjanki@yahoo.com

Willem Joeheo (game warden)
LBB/NB Nature Conservation Division
Cornelis Jongbawstraat 12
Paramaribo Suriname

Sahieda Joemratie (small mammals)
Amazon Conservation Team - Suriname
Nickeriestraat 4
Paramaribo Suriname
s.joemratie@actsuriname.org

Vanessa Kadosoe (aquatic beetles)
Anton de Kom University of Suriname
Leysweg 9
Paramaribo Suriname
vanessakadosoe@gmail.com

Cindyrella Kasanpawiro (reptiles and amphibians)
Anton de Kom University of Suriname
Leysweg 9
Paramaribo Suriname
cindyrellak@gmail.com

Gwendolyn Landburg, M.Sc. (water quality)
Environmental Department
National Zoological Collection of Suriname/Center for
Environmental Research
Anton de Kom University of Suriname
Leysweg 9
Paramaribo Suriname
g.landburg@uvs.edu

Trond H. Larsen, Ph.D. (dung beetles)
Rapid Assessment Program
Conservation International
2011 Crystal Drive, Suite 500
Arlington VA 22202 USA
t.larsen@conservation.org

Burton K. Lim, Ph.D. (small mammals)
Department of Natural History
Royal Ontario Museum
100 Queen's Park
Toronto, ON M5S 2C6 Canada
burtonl@rom.on.ca

Fabian Lingaard (game warden)
LBB/NB Nature Conservation Division
Cornelis Jongbawstraat 12
Paramaribo Suriname
apache-23@live.com

Angelique Mackintosh (large mammals)
Anton de Kom University of Suriname
Leysweg 86
Paramaribo Suriname
angie_mackintosh@hotmail.com

Piotr Naskrecki, Ph.D. (katydids and grasshoppers)
Museum of Comparative Zoology
Harvard University
26 Oxford Street
Cambridge MA 02138 USA
pnaskrecki@oeb.harvard.edu

Sheinh A. Oedeppe (park ranger)
Amazon Conservation Team
Kwamalasamutu Suriname
info@actsuriname.org

Brian J. O'Shea, Ph.D. (birds, logistics coordinator)
Research and Collections
North Carolina Museum of Natural Sciences
11 W. Jones St.
Raleigh NC 27601 USA
brian.oshea@ncdenr.gov

Paul E. Ouboter, Ph.D. (reptiles and amphibians)
National Zoological Collection of Suriname
Anton de Kom University of Suriname
Leysweg 9
Paramaribo Suriname
p.ouboter@uvs.edu

Napoti Pantodina (park ranger)
Amazon Conservation Team
Kwamalasamutu Suriname
info@actsuriname.org

Esteban Payán, Ph.D. (large mammals)
Panthera
Calle 93Bis, #19-40, Oficina 206
Bogotá Colombia
epayan@panthera.org

Martino Piqué (fishes)
I.O.L (Advanced Teachers College) Biology
Paramaribo, Suriname
mar_orpheo@yahoo.com

Serano Ramcharan (birds)
STINASU
Cornelis Jongbawstraat 14
Paramaribo Suriname
sbirdwatching@gmail.com

Jonathang Sapa (park ranger)
Amazon Conservation Team
Kwamalasamutu Suriname
info@actsuriname.org

Teddi Shikoei (park ranger)
Amazon Conservation Team
Kwamalasamutu Suriname
info@actsuriname.org

Andrew E. Z. Short, Ph.D. (aquatic beetles)
Division of Entomology, Biodiversity Institute
Department of Ecology and Evolutionary Biology
1501 Crestline Dr, Suite 140, University of Kansas
Lawrence KS 66045 USA
aezshort@ku.edu

Philip W. Willink, Ph.D. (fishes)
Division of Fishes
The Field Museum
1400 S. Lake Shore Dr.
Chicago IL 60605 USA
p.willink@fieldmuseum.org

Kenneth Wan Tong You (fishes)
National Zoological Collection of Suriname
Anton de Kom University of Suriname
Leysweg 9
Paramaribo Suriname
kgl_wan@hotmail.com

CONSERVATION INTERNATIONAL SURINAME

Conservation International Suriname (CI-Suriname) is a non-profit, non-governmental organization established in Suriname in 1992. CI-Suriname's present focus is centered on the creation and improved management of protected areas, as well as ecotourism development, all underpinned by research, policy support and capacity building of Government institutions, local communities and academia. In Kwamalasamutu, we have been supporting the development and management of a community ecolodge and the wildlife sanctuary created in 2005.

CI-Suriname works in the Kwamalasamutu area with Stichting Meu (Meu Foundation). This is the local development organization, established by the people of Kwamalasamutu, whose responsibility includes the management of the community ecolodge and the Werehpai/Iwana Saamu sanctuary.

Conservation International Suriname
Kromme Elleboogstraat no. 20
Paramaribo
Suriname
Tel: 597-421305
Fax: 597-421172
Web: www.ci-suriname.org

CONSERVATION INTERNATIONAL

Conservation International (CI) is an international, nonprofit organization based in Arlington, VA. CI believes that the Earth's natural heritage must be maintained if future generations are to thrive spiritually, culturally and economically. Building upon a strong foundation of science, partnership and field demonstration, CI empowers societies to responsibly and sustainably care for nature, our global biodiversity, for the well-being of humanity.

Conservation International
2011 Crystal Drive, Suite 500
Arlington, VA 22202
Tel: (703) 341-2400
Web: www.conservation.org

ANTON DE KOM UNIVERSITY OF SURINAME

Anton de Kom University of Suriname was founded on 1 November 1968 and offers studies in the fields of social, technological and medical sciences. There are five research centers conducting research and rendering services to the community. The Center for Agricultural Research (CELOS) is promoting agricultural scientific education at the faculty of Technological Sciences, Institute for Applied Technology (INTEC), Biomedical Research Institute, Institute for Development Planning and Management (IDPM), Institute for Research in Social Sciences (IMWO), the Library of ADEK, University Computer Center (UCC), National Zoological Collection (NZCS) and National Herbarium of Suriname (BBS).

The primary goals of the NZCS and BBS are to develop an overview of the flora and fauna of Suriname, and build a reference collection for scientific and educational purposes. The NZCS also conducts research on the biology, ecology, and distribution of certain animal species or on the composition and status of certain ecosystems.

Anton de Kom University of Suriname
Universiteitscomplex / Leysweg 86
P.O. Box 9212
Paramaribo
Suriname
Web: www.uvs.edu

National Zoological Collection of Suriname (NZCS)
Universiteitscomplex / Leysweg 9
Building # 17
P.O. Box 9212
Paramaribo
Suriname
Tel: 597-494756
Email: nzcs@uvs.edu

National Herbarium of Suriname (BBS)
Universiteitscomplex / Leysweg
Paramaribo
Suriname
Tel: 597-465558
Email: bbs@uvs.edu

AMAZON CONSERVATION TEAM

The Amazon Conservation Team (ACT) was founded in 1996 by ethnobotanist Mark Plotkin and conservationist Liliana Madrigal. ACT is a 501 (c) (3) nonprofit organization supported by individuals, private foundations, and government grants. ACT achieves its conservation successes by working in true partnership with indigenous peoples of the Amazon to protect both their cultures and their ancestral lands - some of the largest tracts of both pristine and sustainably managed rainforest on the planet. Our work begins in the forest — with the people who actually live there — and then expands to the state, national, regional and international level.

Amazon Conservation Team
4211 N. Fairfax Dr.
Arlington VA 22203
Tel: (703) 522-4684
Email: info@amazonteam.org
Web: www.amazonteam.org

ALCOA FOUNDATION

Alcoa Foundation is one of the largest corporate foundations in the U.S. Founded more than 50 years ago, Alcoa Foundation has invested more than US $530 million since 1952. In 2010, Alcoa and Alcoa Foundation contributed $36.8 million — $19.9 from the Foundation — to nonprofit organizations throughout the world, focusing on Environment, Empowerment, Education and Sustainable Design. Through this work, Alcoa seeks to promote environmental stewardship, prepare tomorrow's leaders and enable economic and social sustainability.

The work of Alcoa Foundation is further enhanced by Alcoa's thousands of employee volunteers, who in 2010 gave more than 720,000 service hours — the equivalent of 350 people working full time. In 2010, a record 49 percent of Alcoans took part in the company's signature Month of Service across 23 countries. Through nearly 1,000 activities, Alcoans reached 59,000 children, served 17,000 meals, planted 16,000 trees and supported 3,000 nonprofit organizations.

The Alcoa Foundation can be reached at http://www.alcoa.com/global/en/community/foundation/contact.asp

SURINAME FORESTRY SERVICE (LBB/SBB)

The Forestry Service (LBB) within the Ministry of Physical Planning, Land and Forest Management is responsible for forest management and nature conservation. One of its agencies is the Nature Conservation Department, which is assigned the task of managing all protected areas. As the CITES authority in Suriname, it also issues all permits for research and for the export of species, as well as enforcement within the framework of the laws on hunting and wildlife. The second agency is the Foundation for Forest Management and Production Control (SBB), which is mandated to manage production forests and is responsible for supervision and control of all logging.

PANTHERA

Panthera's mission is to ensure the future of wild cats through scientific leadership and global conservation action. Panthera develops, implements, and oversees range-wide species conservation strategies for the world's largest, most imperiled cats—tigers, lions, jaguars and snow leopards. Through partnerships with local and international NGO's, scientific institutions, and government agencies we apply conservation actions driven to secure key habitat and vital corridors for threatened species. The Jaguar Corridor Initiative is a unique strategy for range-wide long term conservation of the jaguar, as well as vast landscapes and ecosystem functions. In Northern South America we have secured kilometers of land designated as jaguar corridor, strengthened and created new protected areas and work hand-in-hand with our on the ground allies, such as ranchers and indigenous communities.

Panthera USA
8 West 40th St, 18th floor
New York NY 10018
Tel. (646) 786-0400
Fax (646) 786-0401
Web: www.panthera.org

STINASU

The Foundation for Nature Conservation in Suriname (STINASU) was founded in June 1969 and contributes to the protection of Suriname's natural resources and cultural heritage by supporting local and international partnerships in the fields of scientific research, nature education and ecotourism.

Stichting Natuurbehoud Suriname (STINASU)
Cornelis Jongbawstraat 14
Paramaribo
Suriname
Tel: (597) 427102; 427103; 421850; 421683
Fax: (597) 421850
Web: www.stinasu.sr

Acknowledgments

The RAP team extends our deepest gratitude to Granman Ashonko Alalaparoe and the people of Kwamalasamutu for granting us permission to study the biodiversity of their lands. We especially wish to thank John Rudolph, Mopi, Kupias, Koita, Dennis, and Johnny for their leadership in the field. We also thank Ronald, Valentino, George, Metusala, Javan, Fredi, Markus, Ranise, Susa, Paulino, Aasho, Peeri, Keresen and Marai for invaluable field and camp assistance.

We thank Natascha Veninga of CI-Suriname who offered unfailing support and logistical assistance at all stages of the RAP. Annette Tjon Sie Fat and Theo Sno also provided essential support and guidance.

The staff of Gum Air ensured safe transportation of our personnel, equipment, and supplies, and was always able to accommodate the diverse needs of our large team. We thank Maikel Schenkers and Erwin Neles of MediaVision for capturing the excitement and results of the RAP survey on film. We are extremely grateful to our wonderful cook, Gracia Simeon, and her assistant, Chagall Veira, for feeding us during the RAP survey.

We are extremely grateful to the personnel of LBB/NB (Nature Conservation Division), especially Claudine Sakimin, for support in granting us the proper research and export permits. We thank the National Zoological Collection of Suriname and the National Herbarium of Suriname (Anton de Kom University of Suriname) for assistance with the permit application process.

The RAP botanical team gratefully acknowledges the support of the Suriname Forest Control Foundation (SBB), CELOS, and the National Herbarium of Suriname (BBS). For assistance with the identification of plants, we thank Ben Torke, Sir Ghillean Prance, Lars Chatrou, Marion Jansen-Jacobs, Paul Maas, and Tinde van Andel.

The aquatic beetle team is extremely grateful for the many groups and individuals that facilitated this fieldwork. Several specialists provided help with some of the species IDs, and they are all warmly thanked: Dr. Philip Perkins (Harvard University; Hydraenidae), Dr. Kelly Miller (University of New Mexico; Dytiscidae in part), Crystal Maier (University of Kansas; Dryopoidea) and Dr. Martin Fikáček (Czech National Collection; Sphaeridiinae). Taro Eldredge and Sarah Schmits (both University of Kansas) assisted with habitus photographs and database management. We would also like to thank University of Kansas undergraduates Brad Schmidt, Clay McIntosh, and Frazier Graham for assistance with specimen processing. We thank three anonymous reviewers for comments on the chapter manuscript.

The fishes team is indebted to the people of Kwamalasamutu for acting as guides, safely piloting the boats up and down rapids, talking about their fishing experiences, and generously sharing their knowledge of the local rivers. Many people helped us catch specimens, including A. Short, P. Naskrecki, B. O'Shea, E. Payan, E. Neles, T. Larsen, C. Kasanpawiro, A. Mackintosh, H. Gould, George, and Valentino. We thank J. Mol for answering questions about the fishes and hydrology of the region.

The ornithology team thanks Greg Budney and the curatorial staff of the Macaulay Library for the generous loan of recording equipment used on this survey. We also thank John Mittermeier for sharing information from the Yale/Peabody expedition to Werehpai in 2006.

The ant team thanks the entire RAP team for help in collecting ant samples, especially Krisna Gajapersad, Angelique Mackintosh, Chequita Bhikhi, and Willem Joehoe. We also thank Jeffrey Sosa-Calvo (Smithsonian Institution) for help with identifying the ants.

Special thanks to the Amazon Conservation Team (ACT) rangers, to staff from the LBB/NB Nature Conservation Division and SBB/Suriname Forest Service, and to the Surinamese students for joining the RAP team, for providing exceptional assistance to the researchers, and for sharing their knowledge with the team.

We thank Donnell Roy from CI's Center for Environmental Leadership in Business (CELB) and Lisa Famolare from CI's South America Field Division for their help in obtaining funding for this work and for their collaborative support of biodiversity exploration in Suriname.

We thank the Alcoa Foundation for their generous financial support of this RAP survey and for their continued support of conservation and scientific capacity building in Suriname.

Report at a Glance

DATES OF RAP SURVEY

August 18–September 8, 2010

DESCRIPTION OF RAP SURVEY SITES

The Kwamalasamutu region refers to the area of lowland tropical forest surrounding the Trio settlement of Kwamalasamutu. At a minimum, it encompasses the eastern portion of the upper Corantijn watershed, or an area extending from the village south to the Brazilian border, east to the Sipaliwini savanna, north to the Eilerts de Haan and Wilhelmina mountains, and west to the Upper Corantijn River. This vast area is sparsely populated, and its biota is poorly known relative to central and eastern Suriname. The elevation of the region is mostly between 200–400 meters (higher in the south along the Brazilian border), but scattered granitic formations to the north and east of Kwamalasamutu approach 800 m. The region is entirely forested. The RAP team worked around three study sites on the Kutari and Sipaliwini Rivers, each accessible within a day's travel by boat from Kwamalasamutu. The fish and water quality teams also sampled along waterways between our camps. At all sites, the predominant terrestrial habitat was tall forest on both well-drained and seasonally inundated soils.

REASONS FOR THE RAP SURVEY

In 2000, a cave with extensive petroglyphs (Werehpai) was discovered near the village of Kwamalasamutu. Shortly thereafter, the community established the Werehpai/Iwana Samu Sanctuary to serve as an ecotourism site and game reserve, both to generate income for the community and to protect populations of animals upon which the people of Kwamalasamutu depend for food. Conservation International–Suriname has since been working with the community and several donor agencies to establish infrastructure and maintain the sanctuary.

The purpose of this RAP survey was to establish baseline information on the region's biodiversity to inform ecotourism and future monitoring efforts, focusing on Werehpai and the surrounding region. We sought especially to gather information on plant and animal species important to the Trio people, and to provide recommendations that will support sustainable harvest and management practices. The overall goal was to bring together the knowledge and expertise of local people with scientific knowledge to study and plan for monitoring of biological and cultural resources of the Kwamalasamutu region.

MAJOR RESULTS

The RAP team found the Kwamalasamutu region to harbor a rich biodiversity, with few signs of ecosystem degradation. However, there were indications that hunting and fishing pressure have affected the local abundance of some large-bodied mammals, birds, reptiles, and fishes. There was also some evidence of mercury contamination in the rivers, although levels of mercury were considerably lower than has been recorded in watersheds where extensive gold mining occurs. The majority of species found were typical of lowland forests of the Guiana Shield. At least 46 species were new to science, and due to the limited extent of sampling, it is highly likely that many more undescribed species exist in the region.

Number of species recorded

Plants	>240
Ants	>100
Aquatic Beetles	144
Dung Beetles	94
Dragonflies and Damselflies	94
Katydids and Grasshoppers	78
Fishes	99
Reptiles and Amphibians	78
Birds	327
Small Mammals	38
Large Mammals	29

Number of species new to science

Aquatic Beetles	16–26
Dung Beetles	10–14
Dragonflies and Damselflies	4
Katydids and Grasshoppers	7
Fishes	8
Reptiles and Amphibians	1

New records for Suriname

Plants	8
Aquatic Beetles	45
Dung Beetles	5
Dragonflies and Damselflies	14
Katydids and Grasshoppers	29
Fishes	2
Reptiles and Amphibians	2
Birds	4
Small Mammals	2

CONSERVATION RECOMMENDATIONS

The ecosystems of the Kwamalasamutu region face few immediate threats; however, they should be managed to ensure that key ecological processes are not disrupted through contamination of watercourses, large-scale resource exploitation, or depletion of animal populations. We particularly recommend that small-scale gold mining activities be aggressively discouraged in the region. A water quality monitoring program should be implemented by the community to detect contamination from any mining activities. Fifteen species listed on the IUCN Red List of Threatened Species (IUCN 2011) were encountered during the survey. Populations of game animals and fishes should be managed through means best suited to the interests and needs of the community, preferably through a network of reserves with limited hunting seasons for particular species. Domesticated animals should be encouraged as an alternative protein source.

The Kwamalasamutu area's pristine nature, high diversity of birds and other taxa, and abundance of large mammals, including jaguar and ocelot, make it ideal for ecotourism. Tourism should be promoted and supported as a means of protecting the wildlife of the area and of providing employment and livelihood for many people of Kwamalasamutu.

Maps and Photos

Map 1: Werehpai and Iwana Samu Area

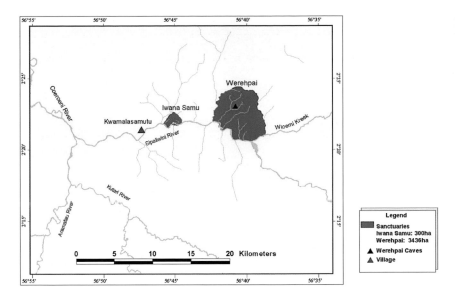

Conservation International Suriname, May 2007

Map 2: Protected Area Werehpai/Iwana Samu

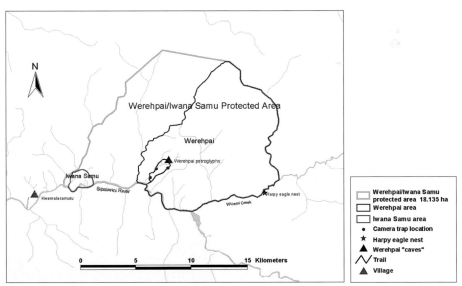

Conservation International Suriname, Nov 2007

Map 3: RAP survey sites

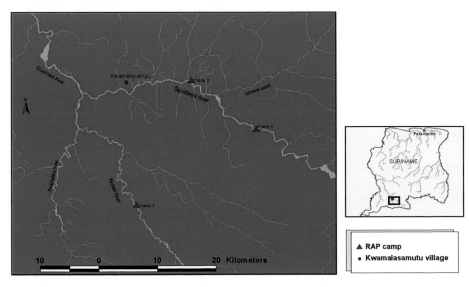

Conservation International Suriname, 2010

The RAP team at the Werehpai base camp. (P. Naskrecki)

The survey site along the Kutari River was characterized by large areas of swamp forest and seasonally inundated forest. (T. Larsen)

Understory palms dominated in some parts of the forest, such as this area at the Werehpai survey site. (T. Larsen)

Piotr Naskrecki photographs the caves in the Werehpai/Iwana Samu Sanctuary. These caves are likely to have been used by humans for at least 5,000 years. (T. Larsen)

In 2000, more than 313 petroglyphs and shards of pottery were discovered in the caves, and the petroglyphs are estimated to be over 3,000 years old. (P. Naskrecki)

A flower of *Cochlospermum orinocense*. (C. Bhikhi)

[above] *Helosis cayennensis* is a parasitic plant that does not produce its own leaves for photosynthesis, but instead sucks nutrients from the roots of trees. (T. Larsen)

Tabernaemontana heterophylla is a plant that contains compounds used to treat dementia and improve memory. Many plants in the Kwamalasamutu Region have potential to yield new medicine and other biotechnological discoveries. (C. Bhikhi)

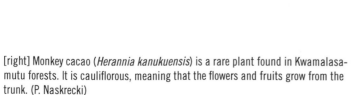

[right] Monkey cacao (*Herannia kanukuensis*) is a rare plant found in Kwamalasa-mutu forests. It is cauliflorous, meaning that the flowers and fruits grow from the trunk. (P. Naskrecki)

Male and female damselflies (*Argia* sp. 1) in tandem at Iwana Samu.

Male of *Argia* sp. 2 at Werehpai Camp. Both species pictured here (*Argia* sp. 1 and 2) are new to science, as are two other *Argia* species found during this RAP survey. *Argia* is the most speciose damselfly genus in the New World. (N. von Ellenrieder)

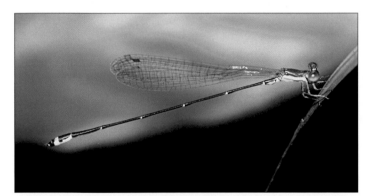

Male damselfly (*Phasmoneura exigua*) at a forest swamp in Kutari Camp. (N. von Ellenrieder)

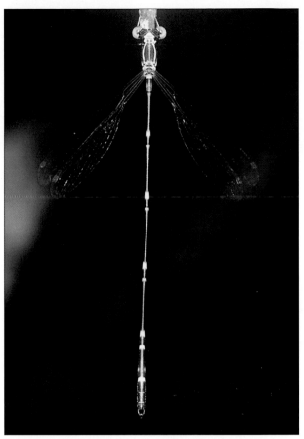

Until now, *Perilestes gracillimus* was known from only two records from Amazonian Peru and Brazil. The female ovipositor is strong and likely used to lay eggs in hard substrates, including bark of twigs. Known larvae in this genus live among dead leaf litter. (N. von Ellenrieder)

Hydrophilus smaragdinus, found in forest pools, is among the largest aquatic beetles in South America. Adults are scavengers of detritus, while the larvae are voracious predators of insects and small fish. (P. Naskrecki)

[left] Adults of *Inpabasis rosea*, an elusive damselfly that has never been photographed until now, can be seen perching on leaves at sunny spots in forest swamps. Larvae and breeding habitat are still unknown, but males were observed guarding small muddy depressions where they most likely reproduce and where females probably lay their eggs. (N. von Ellenrieder)

Oxysternon festivum is a large, brightly colored dung beetle. (T. Larsen)

Phanaeus chalcomelas males use their horns as weapons to fight over potential mates. (T. Larsen)

Deltochilum valgum is a highly unusual dung beetle species that is specialized exclusively to prey on live millipedes. Its elongated hind legs are used to wrap around the body of millipedes that are much larger than the beetle. This behavior was unknown until two years ago. (T. Larsen)

Loboscelis bacatus is a spectacular conehead katydid, previously known only from Amazonian Peru, but found during the Kwamalasamutu RAP in southern Suriname, significantly extending its known range. (P. Naskrecki)

[left] *Coprophanaeus lancifer* is the largest of all Neotropical dung beetle species. Both sexes possess long horns that are used during intrasexual battles. This species can rapidly bury an animal carcass as large as a pig. (P. Naskrecki)

Eubliastes adustus, a sylvan katydid previously known only from Ecuador (P. Naskrecki)

Vestria sp. n. This new, yet unnamed species of conehead katydid was discovered during the Kwamalasamutu RAP. (P. Naskrecki)

Gigantiops destructor – the Jumping Ant – is a large black ant common on the forest floor in the Werehpai area. *G. destructor* has excellent vision, and solitary workers of this species can be seen during the day as they scurry across the forest floor looking for prey. (P. Naskrecki)

Camponotus sericeiventris – the Carpenter Ant – is one of the largest and shiniest ants in the forest. This soldier may guard the nest located in trunks or large branches of large live trees while the workers scavenge for food during the day. (T. Larsen)

Cephalotes atratus – the Turtle Ant, or Gliding Ant, lives high up in the tree canopy and can glide back to their home tree if they fall. (P. Naskrecki)

Daceton armigerum – the Canopy Ant – is a beautiful golden-colored ant that lives high in the canopy of trees near the Werehpai caves. (P. Naskrecki)

RAP researcher Phil Willink processes fish specimens, including a Black Piranha (*Serrasalmus rhombeus*). The Black Piranha is a top-level aquatic predator – this specimen was 41 centimeters long, and weighed 3 kilograms. (P. Naskrecki)

This fish species, *Pterodoras* aff. *granulosus*, is probably new to science. (P. Willink)

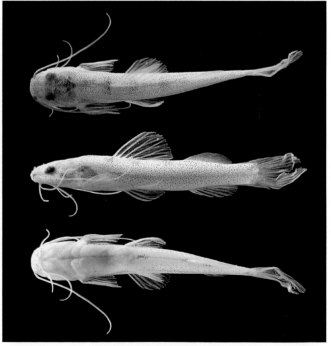

Armored catfish (*Pseudancistrus corantijniensis*) are associated with rocky bottoms of fast flowing rivers and streams. This species uses its spoon-shaped teeth to scrape algae off of logs and rocks. (P. Naskrecki)

Another fish species, *Imparfinis* aff. *stictonotus*, collected during the RAP survey which may be new to science. (P. Willink)

The Dyeing poison frog (*Dendrobates tinctorius*) has highly toxic skin, and the frog advertises its noxious properties with its very noticeable colors. The frogs obtain their toxins from ants, on which they feed. (P. Naskrecki)

The Suriname horned frog (*Ceratophrys cornuta*) is a voracious sit-and-wait predator. It has an exceptionally wide mouth, which allows it to swallow prey that is nearly as large as its own body, including mice and other frogs. (T. Larsen)

Three-striped poison dart frog (*Ameerega trivittata*) with tadpoles on the back. Adults of many poison dart frog species transport their young from one body of water to another as the tadpoles feed and develop. (T. Larsen)

The Amazon tree boa (*Corallus hortulanus*) is a small constrictor, which feeds primarily on rodents and birds. (P. Naskrecki)

Gonatodes humeralis is a brightly colored dwarf gecko that is active during the day. (T. Larsen)

RAP researcher Burton Lim takes a fruit bat (*Artibeus planirostris*) from the mist net. (P. Naskrecki)

Four species of bats from the RAP. Clockwise from upper left: 1) large fruit-eating bat (*Artibeus planirostris*), 2) nectar-feeding bat (*Lonchophylla thomasi*), 3) sword-nosed bat (*Lonchorhina inusitata*), and 4) moustached bat (*Pteronotus parnellii*). (B. Lim)

Four species of rodents from the RAP. Clockwise from upper left: 1) spiny mouse (*Neacomys* sp.), 2) McConnell's rice rat (*Euryoryzomys macconnelli*), 3) spiny rat (*Proechimys* sp.), and 4) terrestrial rice rat (*Hylaeamys megacephalus*). (Photos 1 & 2 by B. Lim, 3 & 4 by E. Neles)

The Jaguar (*Panthera onca*), also known as Kaikui, is the largest cat of the Americas. Jaguars eat a variety of animals such as peccaries, tapirs, cattle, and deer. Jaguars swim well and their habitats range from rainforest to dry forest. Jaguars are rare in many places due to hunting and the fur trade. (K. Gajapersad)

The Ocelot (*Leopardus pardalis*) occurs in forested landscapes throughout the Neotropics, but is active mostly at night, and is therefore rarely seen. Camera trap images gathered during the RAP, such as this one from the Kutari site, suggest that this species is common in the Kwamalasamutu region. (K. Gajapersad)

Despite its large size and heavy body armature, the Giant Armadillo (*Priodontes maximus*), known as Morainmë in Trio, is a gentle animal, feeding primarily on termites and ants, which it digs out from underground nests using its huge claws. Giant armadillos are rarely seen due to their nocturnal habits, but their huge burrows are a common sight in the forests of Kwamalasamutu. This species is declining across its range in South America, primarily as a result of excessive hunting and habitat loss. (K. Gajapersad)

The Paca or Kurimau (*Cuniculus paca*) is a large caviomorph rodent of Neotropical rainforests that is highly prized for food across its vast range. They remain common in forested landscapes like the Kwamalasamutu region, where the human population density is relatively low. The Paca's diet includes fallen fruit, and they are important seed dispersers for many rainforest trees. (K. Gajapersad)

The Collared peccary (*Pecari tajacu*) is the smaller of the two species of peccaries in the Kwamalasamutu region, and is one of the preferred food animals for the Trio people. (K. Gajapersad)

The Black Curassow (*Crax alector*), known in Trio as ohko, is a large bird, found along rivers and in forests across most of the Guiana Shield. These birds spend most of their time on the ground, feeding on insects, fruits, and seeds. In the Kwamalasamutu Region they are hunted by the Trio, but this species remains plentiful away from human settlements. (K. Gajapersad)

Iponohto Pisi Serë

AWAINAHTO IMENEHKATOPONPË

August 18 – September 8, 2010

Ampo imenehkatoponpë sere, Kwamalasamutu nai konopopijan itu kato, irë nai Tareno ipata Kwamalasamutu. Irë nai Corantijn tuna iwehtihatopo. Irë nai zide weije Kwamalasamutu pata pëe. Wei wëehto weije nai Sipaliwini ijoi. Ma roord weije nai Eilert de Haan ma Wilhelmina pïiton, weste weije nai Guayana inono intakato Suriname pëe. Serë nai mono nono wïtoto waken kunme irë inonopo. Irë nai wïtoto menekaewankiërë kure Central ma ooste weije mare. Irë pata nai kawë tunaimë epae 200 – 400 meters, karaiwa inono intakato weije kawe irëtëpoeton pïme irëpo, irë nai 800 meters kawae iwekto. Irë nono nai itume reken. Imenekato kinei 3 inonopo, irë nai kuitaritunatae, Sipaliwini itunatae mare. Montoruke nai irë pona wïtëto 1 wei. Irë imenekato tese, kanaton ma tunaton irë imenekane tese, iwënïtokon wararë. Kanputoton wararë kawiëno ituton tïmenekae ijane, nono mare tïmenekae ijane.

ATÏJANME TIMENEKAE

2000 mao ahtao tëpoe totaken imenuhtëpëe tërahtëe ijane maahtaken pata Kwamalasamutu pë. Ireme Kwamalasamutu pontomoja Werehpaeme tekatëe. Ma Iwanasamoe mare tirëe ijane toeriston iwëehtome ijane, irëjanma marë mehparëton tïpïnmae ijane. Irë apo tïrëe ijane karakuri epohtome pata akoronmahtome ijane. Irë mare tïpïnmae ijane otikonme iweike. Ma CI tonmarë akërëne tëse ma karakuri entuton mare kure tinonokon tïrïtome ijane.

Irë nono imenekato tïrëe ijane toeriston iwëehtome irëpona ma karakuri epohtome mare ijane. Irëme Werehpai inonotipïnmae ijane. Irëjanme onipekenton timenehkapoe ijane ituton marë mehparëton irasaton tarenotomoja iweke.

Irëjame tikurumae ijane teperukenton, tënasehton kure tïritome ijane. Irëjanme sereton tirëe ijane eisaporo tiwipunehtokon tïrïtome ijane Tarenoton akërë. Kure tïnonokon tïrïtome ijane irëpëe karakuri ëpohtome ijane tïpatakonporo.

IRASATON EPORÏHPËTON IRËMAO

Irëme imenekanepëeton kan irëpëe tïjapëkenai irënono. Onipekenton imenekahpëeke ijane, këpëwa wëiwatojanme mehparëton ototon ma kanaton wakensa tese irëpo.

Ma tuna imenekanmahtao ijane irëmao kuweki apo tërahtëe ijane tunahkao. Ma irëme imenehkaneton tïikae atïtome koutupatapo waken tuna iwëtanmëpo kuweki ma serëpo kuweki pije. Ma imenekatuwe ijane senpo irëmao Guyana iwehto aporopa tëne ijane. Ma imenekatomao irëmao tapïme tëne ijane tiwamekatohton serë inonopo.

Okinpëken Ekaton

Ituton	>240
Situton	>100
Ëënëton	144
Pëmuton	94
Pïmokokoton	94
Sunariton	78
Kanaaton	99
Ëkeiton/Përëruton	78
Tonoroton	327
Mëhparë akïiton	38
Inunu mëhparëton	29

Ahtare Kainan Ërahtëtoponpë

Ëënëton	16–26
Pëmuton	10–14
Pïmokokoton	4
Sunariton	7
Kanaaton	8
Ëkeiton/Përëruton	1

Kainan eratëhpëton Surinampo

Ituton	8
Ëënëton	45
Pëmuton	5
Pïmokokoton	14
Sunariton	29
Kanaaton	2
Ëkeiton/Përëruton	2
Tonoroton	4
Mëhparë akïiton	2

TÏPÏNMAINPË KURE TIRËENME

Kwamalasamutu inonopo antinaosa teese. Këpëewa irë nono tikurunmaenme nai, irëmao tuna ikëhrëmaeto nehtan. Ma mëhparëton ma ototon mare kure nehtan. Irëme koutupëkenton sameken takamainme nehtan irënonopo. Irëme irë tuna timenkapore nehtan oroko wehto wararë. Irëme irëpatapontomoro nikurunmatan tïnonokon. Irëmao wëeiwato tïpatoro nehtan mëhparëton wëtohpë. Irëjanme kurairuton, kuusiton arimikato kure nehtan ijane, irëmao wëeiwato waken nehtan.

Kwamalasamutu inono nai maa pata. Okïhnpëkenton, tonoroton, pimokokoton, tëpëriken pisiton, tïkapïpëhkenton ma mëhparëton marë. Ma ëpëton kaikuiton marë. Irë nai kure toerestomoja. Toereston amohtëtome ijane. Irëjanme kure tïpïne tïrëeinme mehparëton, ototon, ituton mare. Irë mare nai Tarënoton aeneme iwehtohkon mare tïpatakonpo Kwamalasamutupo.

Rapportage in Vogelvlucht

DATUM VAN HET RAP-ONDERZOEK

18 augustus – 8 september 2010

BESCHRIJVING VAN DE RAP-ONDERZOEKSLOCATIES

Het gebied waarnaar wordt vewezen is het laagland tropisch bos dat het Trio-dorp Kwamalasamutu omringt, oftewel het oostelijke deel van het stroomgebied van de boven-Corantijn. Dit is het gebied ten zuiden van Kwamalasamutu naar de grens met Brazilie, ten oosten naar de Sipaliwini-savanna, ten noorden naar het Eilerts de Haangebergte en het Wilhelminagebergte, en ten westen tot aan de Boven-Corantijn. Dit uitgestrekte gebied is dunbevolkt en de fauna en flora zijn niet erg bekend in vergelijking met centraal en oost-Suriname. Het gebied ligt grotendeels op een hoogte van 200–400 meter (hoger in het zuiden langs de grens met Brazilië), maar hier en daar bevinden zich granietachtige formaties naar het noorden en het oosten, ongeveer op een afstand van 800 meter van de toegangsweg naar het dorp Kwamalasamutu. Het hele gebied is bedekt met bos. Het RAP-team verrichtte werkzaamheden in drie onderzoekslocaties aan de Kutari- en de Sipaliwinirivier, die elk gemakkelijk toegankelijk waren en binnen een dag met de boot vanuit Kwamalasamutu konden worden bereikt. De teams belast met het onderzoeken van de vis- en waterkwaliteit trok ook monsters langs de kreken en rivieren tussen onze kampen. Op alle locaties was de voornaamste terrestrische habitat (ofwel leefgebied op het land), hoog bos, zowel op grond met goede afwatering als op grond die, afhankelijk van het seizoen, onder water komt.

REDENEN VOOR HET RAP-ONDERZOEK

In 2000 werd er nabij het dorp Kwamalasamutu een grot ontdekt met uitgebreide rotstekeningen (Werehpai). Kort daarna stelde de leefgemeenschap het Werehpai/Iwana Saamu beschermd gebied in. Met het instellen van dit beschermd gebied wilde men het ecotoerisme bevorderen en inkomsten genereren voor de gemeenschap. Maar wilde men de populaties van dieren, waarvan de inwoners van Kwamalasamutu afhankelijk zijn voor hun voedsel, beschermen. Conservation International Suriname werkt samen met de gemeenschap van Kwamalasamutu en met verschillende donorinstanties om de nodige infrastructuur te ontwikkelen en het beschermd gebied in stand te houden.

Het doel van dit RAP-onderzoek was om informatie over de biodiversiteit van de regio vast te stellen, zodat personen en/of instanties die belast zijn met ecotoerisme, de informatie kunnen gebruiken. Maar de RAP-resultaten zijn ook belangrijk voor monitoring in de toekomst. Het onderzoek concentreerde zich op het gebied van Werehpai en omgeving. We hebben geprobeerd om voornamelijk informatie te verzamelen over plant- en diersoorten die van belang zijn voor de Trio-bevolking en om aanbevelingen te doen voor duurzame houtwinnings- en beheerspraktijken. Het algemene doel was om kennis en expertise van de lokale bevolking te combineren met wetenschappelijke kennis en zodoende de biologische en culturele rijkdommen van het Kwamalasamutugebied vast te leggen en te behouden.

VOORNAAMSTE RESULTATEN

Het RAP-team ontdekte dat het Kwamalasamutugebied een rijke biodiversiteit heeft, met weinig tekenen van degradatie van de ecosystemen. Er waren echter wel wat indicaties die erop wezen dat de druk van jacht en visvangst plaatselijk de overvloed aan enkele grote zoogdieren, vogels, reptielen en vissen heeft aangetast. Er waren ook sporen van kwikverontreiniging in de rivieren, hoewel de kwikgehalten aanzienlijk lager waren dan in stroomgebieden waar er veel goudwinning plaatsvindt. De meeste biologische soorten die zijn aangetroffen, waren kenmerkend voor het laaglandbos van het Guianaschild. Er waren ook enkele soorten die nieuw zijn voor de wetenschap. Door de beperkte omvang van het nemen van monsters is het zeer waarschijnlijk dat er veel meer biologische soorten in het gebied voorkomen waarover er geen documentatie bestaat.

Aantal soorten dat is vastgelegd

Planten	>240
Mieren	>100
Waterkevers	144
Mestkevers	94
Libellen en waterjuffers	94
Sabelsprinkhanen en sprinkhanen	78
Vissen	99
Reptielen en amfibieën	78
Vogels	327
Kleine zoogdieren	38
Grote zoogdieren	29

Aantal soorten dat nieuw is voor de wetenschap

Waterkevers	16–26
Mestkevers	10–14
Libellen en waterjuffers	4
Sabelsprinkhanen en sprinkhanen	7
Vissen	8
Reptielen en amfibieën	1

Nieuwe soorten voor Suriname

Planten	8
Waterkevers	45
Mestkevers	5
Libellen en waterjuffers	14
Sabelsprinkhanen en sprinkhanen	29
Vissen	2
Reptielen en amfibieën	2
Vogels	4
Kleine zoogdieren	2

AANBEVELINGEN

Het Kwamalasamutugebied wordt wel bedreigd door vervuiling van rivier en kreken, mogelijke grootschalige exploitatie van hulpbronnen en het uitdunnen van dierenpopulaties. Deze bedreigingen zullen aangepakt moeten worden om te voorkomen dat belangrijke ecologische processen ontwricht raken, zoals in andere gebieden. Vooral kleinschalige goudwinningsactiviteiten moeten ontmoedigd worden in dit gebied, omdat daarmee veel vernietiging van het milieu gepaard gaat. Ook zou een programma kunnen worden opgezet om de waterkwaliteit te controleren, zodat de gemeenschap al in een vroeg stadium verontreiniging kan ontdekken. Ten slotte, is het belangrijk dat de populaties van wilde dieren en vissen worden beheerd om de eiwitvoorziening van de lokale gemeenschap in de toekomst te garanderen. Dit zou, bijvoorbeeld, kunnen door het gebied in te delen in gebieden waar jachtseizoenen gelden voor bepaalde dieren. Maar ook het kweken van dieren in het dorp zou een alternatieve bron van eiwitten kunnen zijn.

De ongerepte natuur in het Kwamalasamutugebied, de enorme diversiteit aan vogels en andere taxonomische groepen, en de overvloed aan grote zoogdieren, waaronder jaguar en ocelot, maakt dit gebied uitermate geschikt voor ecotoerisme. Toerisme dient te worden bevorderd en ondersteund als middel om de in het wild levende dieren van het gebied te behouden, maar ook om werkgelegenheid en een middel van bestaan voor Kwamalasamutu verder te ontewikkelen.

Executive Summary

INTRODUCTION

The Guiana Shield is a vast tropical wilderness covering over 2.2 million square kilometers and encompassing all or part of six South American countries (Hammond 2005). The numerous biomes of the Guiana Shield have fostered the evolution of an exceptionally rich flora and fauna with many endemic species. More than 20,000 species of vascular plants, 1,000 species of birds, and 1,100 species of freshwater fishes are known from the Guiana Shield (Huber and Foster 2003; Hollowell and Reynolds 2005; Vari et al. 2009). The region's tumultuous cultural history and general remoteness from large population centers have effectively limited environmental degradation on a large scale. As a result, much of the Guiana Shield remains forested, presenting an invaluable opportunity to set conservation goals and develop ecologically and socially responsible strategies for resource use (Huber and Foster 2003; Hammond 2005).

Suriname is entirely contained within the Guiana Shield region and is mostly covered by lowland rainforest. Although most of the human population lives on the coastal plain, many Maroon and Amerindian communities are found in the interior—the former mostly along rivers in the eastern half of the country, and the latter primarily in the far southern and western regions. Much of southern and western Suriname is sparsely populated, and wildlife is abundant.

However, the isolation that has protected Suriname's ecosystems, natural resources, and indigenous cultures is coming to an end, and the opportunity to take action to preserve these remarkable resources may soon be gone. Record high commodity prices have encouraged the spread of illegal gold miners from Brazil across the region, spurred potential major hydropower and mining investments, and provided the incentive to press ahead with road and dam projects.

One of the first steps needed to develop conservation and management plans for Suriname is to collect baseline biological and socio-economic data. Suriname currently lacks the scientific capacity to conduct multi-taxa biodiversity field surveys needed to make sound resource management decisions for the country. Increasing Suriname's scientific capacity is critical to ensuring long-term conservation of the country's biodiversity and promoting sustainable development for Suriname's people. This RAP survey was conducted to incorporate national scientific capacity and train local students, scientists and community members in biodiversity assessment and monitoring methods, as RAP has done before in previous surveys in Suriname (Alonso and Berrenstein 2006, Alonso and Mol 2007).

THE KWAMALASAMUTU REGION

The indigenous settlement of Kwamalasamutu (N 02.3561°, W 056.7945°) is situated on the north bank of the Sipaliwini River in southwest Suriname, approximately 10 kilometers upriver from the confluence of the Sipaliwini and Coeroeni Rivers, which together form the main eastern tributary of the Corantijn River flowing north to the Atlantic Ocean (see Map, page 13). The village of Kwamalasamutu was officially created in 1975. Different small nomadic tribes that inhabited the vast forests of the surrounding region had already been brought together by missionaries in the late 1950s and early 1960s in settlements that were initially more upstream along the Sipaliwini River, but as more people came together in a more or less permanent settlement, the need for more water and better hunting lands eventually led them to their current location. Following establishment of the village, the population reached a maximum of more than 2,000 people before slowly decreasing to its present size of approximately 800 (Teunissen and Noordam 2003). The word 'Tareno' is used by the people themselves as a collective term for the different tribes who live together in Kwamalasamutu. The largest of these tribes is the Trio, and the Trio language is the lingua franca of the people. This document therefore uses 'Trio' to refer to the indigenous people of this region. Today, Kwamalasamutu is the political and cultural center for the Trio people of Suriname. Residents of Kwamalasamutu subsist primarily on fish, bushmeat, and a limited variety of food crops, especially cassava, that are cultivated in a large network of shifting plots surrounding the village (Teunissen and Noordam 2003).

The Kwamalasamutu Region is here considered to encompass the eastern portion of the upper Corantijn watershed, or an area extending from the village south to the Brazilian border, east to the Sipaliwini savanna, north to the Eilerts de Haan and Wilhelmina mountains, and west into the area between the Kutari and Upper Corantijn Rivers. This is one of the most remote areas of the Guiana Shield; the nearest roads are far to the south in Brazil, and all travel within the region is by boat or on foot. A wide, grassy airstrip in Kwamalasamutu serves as the principal connection between the region and the coast. The elevation of the region is mostly between 200–400 meters (higher in the south along the Brazilian border), but scattered granitic formations to the north and east of Kwamalasamutu approach 800 m. The region is entirely forested, with few permanent human settlements.

The Kwamalasamutu Region is situated within the Acarai-Tumucumac priority area located in Guyana, Suriname and Brazil, as defined by participants in a conservation priority-setting workshop held in Paramaribo in 2002, and co-sponsored by Conservation International, the Guiana Shield Initiative of the Netherlands Committee for the International Union for the Conservation of Nature (GSI/NC-IUCN), and the Caribbean sub-regional Resource Facility of the United Nations Development Program (Huber and Foster 2003). Participants ranked the Acarai-Tumucumac area of highest biological importance, citing the area's intact habitats and high ecological diversity, but acknowledged that insufficient data existed to inform conservation recommendations for particular taxonomic groups. To fill this knowledge gap, RAP surveys of the Acarai-Tumucumac area have since been undertaken in both Guyana (Alonso et al. 2008) and Brazil (Bernard 2008), but large areas remain unexplored. This is particularly true in Suriname, where previous biological surveys have been concentrated primarily in the eastern and central regions of the country.

The great forests of the Kwamalasamutu Region extend unbroken far into both Guyana and Brazil, and the region's lack of infrastructure ensures that there is little immediate threat of large-scale extractive activities or landscape conversion. However, this can be expected to change in coming decades as the countries of the Guiana Shield continue to expand their economic activities and develop road networks and other trade connections with one another. Of particular concern from a conservation perspective are the region's populations of large-bodied vertebrates, which are sensitive to over-exploitation and habitat alteration, and the integrity of the region's watercourses, which are vulnerable to sedimentation and contamination from small-scale mining activities.

CONSERVATION INTERNATIONAL'S RAPID ASSESSMENT PROGRAM (RAP)

Conservation International's (CI) Rapid Assessment Program (RAP) is a leading world expert in the collection of field data. RAP is an innovative biological inventory program designed to use scientific information to catalyze conservation action. RAP methods are designed to rapidly assess the biodiversity of highly diverse areas and to train local scientists in biodiversity survey techniques. Since 1990, RAP's teams of expert and host-country scientists have conducted 80 terrestrial, freshwater aquatic (AquaRAP), and marine biodiversity surveys and have contributed to building local scientific capacity for scientists in over 30 countries. Biological information from previous RAP surveys have supported the protection of millions of hectares of tropical forest, including the declaration of protected areas in Bolivia, Peru, Ecuador, and Brazil and the identification of biodiversity priorities in numerous countries. Visit https://learning.conservation.org/biosurvey/Pages/default.aspx for more information on RAP and its methodology.

Capacity Building
In 2008, CI's RAP and Suriname programs carried out two mini-training courses in rapid biodiversity assessment methods for 28 Surinamese students and eight international students. The courses were designed to promote interest in biodiversity conservation and provided students with an introduction to biodiversity assessment and its applications to conservation. Both courses were taught by taxonomic experts who covered field methods for assessing the diversity of plants, large mammals, birds, reptiles and amphibians, and a variety of terrestrial insect groups. Basic introduction to the taxonomy of each group was also presented and practiced. The course format included field projects and activities, lectures, and data analysis.

Project Initiation
In 2000, caves with more than 313 petroglyphs and shards of ancient pottery were discovered at Werehpai, near the village of Kwamalasamutu. At the time, Conservation

International-Suriname (CI-Suriname) was working with the people of Kwamalasamutu on a medicinal plant project that was coming to an end, and decided to explore the region and to support the development of eco-tourism. CI-Suriname first supported an archeological study of the petroglyphs by the Smithsonian Institution and the Suriname Museum (Sandoval 2005), which found evidence to show that the petroglyphs are over 3000 years old and that the caves had been used by humans for at least 5000 years. In 2006, CI-Suriname obtained funds to implement two projects to support protection of the petroglyphs and surrounding forest, and to help reduce poverty in Kwamalasamutu by establishing a community owned and operated eco-tourist facility. The first project, funded by the Global Conservation Fund, aimed to develop two sanctuaries or Indigenous Protected Areas (IPAs) around the petroglyphs (see Map 1, page 13), to identify an appropriate legal mechanism for establishing the areas as sanctuaries, and to build community capacity to manage existing protected areas. At the same time, a second project, funded by the Interamerican Development Bank / Japan Fund, established a community tourism lodge in the Iwana Samu protected area. The tourism lodge was created to generate funds required to sustain effective management of the protected areas, following the management plan that was created in the course of the GCF project.

The local foundation, Stichting Meu, was assigned responsibility by the *Pata Entu* (=chief) of Kwamalasamutu for development and management of protected areas. In 2007, the two separate sanctuaries (Iwana Samu and Werehpai) were joined into one protected area (Werehpai/Iwana Samu Protected Area) and placed under management of Stichting Meu. The total area is now ca. 18,000 ha (see Map 2, page 13). The reason given by the village council for joining the two separate areas into one large protected area was ease of management. One large protected area was easier to delineate than two smaller areas, and everyone on the ground could more easily understand the boundaries of a single sanctuary. Bushmeat hunting is prohibited in the Iwana Samu sanctuary to promote sustainable wildlife populations.

CI-Suriname drafted a report recommending that the Werehpai site be proclaimed a national heritage site, and that the government of Suriname apply to UNESCO for World Heritage Status for the site. The report was submitted to the Minister of Education (who is responsible for cultural sites) in early 2006 and again in 2007. The ministerial team promised further action, but nothing has been forthcoming, due in part to the complexity of indigenous land rights issues in Suriname. CI-Suriname has studied the possibility that Indigenous Protected Areas receive official protected status, and has discussed this complex matter with the government. CI-Suriname recommended that lands indicated as Indigenous Protected Areas be officially issued to the village council of Kwamalasamutu under the Forestry Act. The government installed a committee to study indigenous land rights and to make recommendations for amendments to existing laws or the drafting of new legislation. At the present time, a decision on the official status of the Indigenous Protected Areas established under this project has been postponed until the larger national issue of tribal lands is resolved.

The RAP survey of the Kwamalasamutu Region forms an integral part of CI-Suriname's ongoing efforts to assist the people of Kwamalasamutu to strengthen their capacity to manage the sanctuary and promote tourism on their lands.

The RAP survey in 2010 had two principal goals:

1) supply baseline data on the region's biodiversity and water quality to the Trio people of Kwamalasamutu, including recommendations for the management of game and fish populations for long-term viability and information to support eco-tourism

2) provide Surinamese students and young professionals with further training and opportunities to advance their interests in environmental biology. RAP and CI-Suriname are dedicated to continue to work with these and other students to build capacity for conservation within the university student population.

OVERVIEW OF THE RAP SURVEY

The scientific team included scientists from the Anton de Kom University of Suriname, Conservation International, Panthera, the Amazon Conservation Team, the Museum of Comparative Zoology at Harvard University, the Louisiana State University Museum of Natural Science, the Biodiversity Institute at the University of Kansas, the California State Collection of Arthropods, the Field Museum, the Royal Ontario Museum, and the National Herbarium of the Netherlands. The scientists were joined by seven students currently or formerly enrolled at the University of Suriname, many of whom participated on RAP training courses conducted by CI in Suriname in 2008. The RAP team collected data on water quality, plants, and the following groups of animals: ants, aquatic beetles, dung beetles, dragonflies and damselflies, katydids and grasshoppers, fishes, reptiles and amphibians, birds, small mammals, and large mammals.

Survey sites were chosen to maximize the diversity of sampled habitats, with a particular emphasis on areas most likely to be visited by tourist groups. All sites were easily accessible within one day's travel by boat from Kwamalasamutu.

DESCRIPTION OF THE RAP SURVEY SITES

The RAP team surveyed three sites in the Kwamalasamutu region. Only the coordinates of the base camps are given here; most sampling was done within 5–10 kilometers of these camps. Certain groups were sampled in other areas as well (e.g., along rivers between camps); please refer to individual chapters for sampling protocols and localities.

Site 1. Kutari River
N 02° 10'31", W 056° 47' 14"
18–24 August 2010

The first camp was situated on the east bank of the Kutari River, approximately 44 km by river from Kwamalasamutu. The Kutari flows north from its source along the Suriname-Brazil border and joins the Aramatau to form the Coeroeni River; at our camp, the Kutari formed a meandering channel approximately 40 meters wide. The habitat at this site was a mix of terra firme and seasonally inundated forest, with the latter more extensive here than at our other sites. Away from the river the terrain was quite hilly and supported tall terra firme forest; low-lying areas between hills were often swampy and dominated by palms (*Euterpe oleracea*). At least one large patch of tall bamboo (*Guadua* sp.) was found here as well. Approximately six km of trails were cut at this site, and most terrestrial sampling was done along these trails.

Site 2. Sipaliwini River
N 02° 17' 24", W 056° 36' 26"
27 August – 2 September 2010

The second camp was situated on the north bank of the Sipaliwini River, approximately 27 km upriver from Kwamalasamutu. Here the Sipaliwini formed a broader, straighter channel than the Kutari, and contained numerous boulders and rapids. The habitat around this site was primarily tall terra firme forest, with fewer palm swamps and generally less seasonally flooded forest than the Kutari site. The understory contained many spiny palms (*Astrocaryum sciophilum*). In many places, particularly on hilltops, the soil layer was very thin and supported a shorter forest with fewer large-diameter trees. From this site, we were able to access a small granitic outcrop, or inselberg, situated approximately three km from the camp. Many creeks flowed into the Sipaliwini around this site; some of these creeks had steep banks and formed channels up to 15 meters across. At this site, we sampled primarily along the trail to the inselberg, and along a second trail that extended approximately three km northeast of the camp.

Site 3. Werehpai
N 02° 21' 47", W 056° 41' 52"
3–7 September 2010

The third camp was located on the north bank of the Sipaliwini River, approximately 16 km downriver from the second camp. The river here was slightly wider than at the previous site. The camp itself was situated on an abandoned farm, and the habitat immediately surrounding the camp was mostly second growth forest and bamboo with a dense, almost impenetrable understory. Farther away from the camp, the habitat consisted of tall primary terra firme forest, similar to the previous site. However, the soil was deeper and richer in some areas at this site, supporting more large-diameter trees. Most sampling occurred along the well-established 3.5-km trail to the Werehpai caves. No other trails were cut at this site. From this camp, some groups (primarily the fish and

water quality specialists) surveyed Wioemi Creek, a small river that flows into the Sipaliwini approximately five km upriver from Werehpai. Wioemi Creek was much like the Kutari River in many respects, and supported substantial areas of seasonally flooded forest.

OVERVIEW OF RAP RESULTS—GENERAL IMPRESSIONS

The RAP survey team found the Kwamalasamutu region to be highly diverse and in near-pristine ecological condition. At least 1,316 species of plants and animals were identified by the RAP scientists, of which a minimum of 46 species—16 aquatic beetles, ~ten dung beetles, four dragonflies, seven katydids, eight fishes, and one frog—are new to science. Approximately 111 species were recorded for the first time from Suriname, underscoring the magnitude of the country's biodiversity and the need for additional surveys in other unexplored areas of the region.

Fifteen species listed on the IUCN Red List of Threatened Species (IUCN 2011) were encountered during the survey (Table 1). Many of these species play important roles in the forest ecosystem as top predators and dispersers of large seeds; they also include some of the most highly prized animals in the diet of the Trio people. More data are needed

Table 1. Species listed on the IUCN Red List of Threatened Species that were recorded during the Kwamalasamutu RAP survey. Species are listed in ascending order of threat level; those without English names are trees for which standardized English names do not exist. LR/NT: Lower Risk/Near Threatened; NT: Near Threatened; VU: Vulnerable; EN: Endangered; CR: Critically Endangered.

Scientific name	English name	IUCN Red List Status
Minquartia guianensis		LR/NT
Harpia harpyja	Harpy Eagle	NT
Tayassu pecari	White-lipped Peccary	NT
Panthera onca	Jaguar	NT
Cedrela odorata		VU
Corythophora labriculata		VU
Chelonoides denticulata	Yellow-footed Tortoise	VU
Ateles paniscus	Guianan Spider Monkey	VU
Priodontes maximus	Giant Armadillo	VU
Myrmecophaga tridactyla	Giant Anteater	VU
Tapirus terrestris	Brazilian Tapir	VU
Aniba rosaedora		EN
Trichilia surumuensis		EN
Pteronura brasiliensis	Giant Otter	EN
Vouacapoua americana		CR

to inform suitable hunting quotas for these species, but the RAP data provide some baseline information on species distribution and hunting intensity to inform a process to determine sustainable hunting levels.

RAP RESULTS BY TAXONOMIC GROUP

Water Quality

A total of 23 sites were sampled intensively in three major areas: the Kutari River, and two areas of the Sipaliwini River. We measured 13 physico-chemical parameters at each site: pH, dissolved oxygen, conductivity, temperature, alkalinity, total hardness, total phosphate, nitrate, chloride, tannin & lignin, ammonia, turbidity and secci depth. The oxygen content and pH of the Kutari River were lower than those of the Sipaliwini River, probably due to the lack of rapids and the input of organic material from the surrounding forest, particularly after heavy rains, which occurred frequently at the Kutari site. All sites had clear water except the Wioemi Creek, which was very turbid. The parameters measured in the field revealed undisturbed river ecosystems with few negative human impacts. However, high mercury levels were found in both sediment and piscivorous fishes from all sites. Further research is needed to clarify the origin of mercury in these river systems, and we recommend initiating a water quality monitoring program in Kwamalasamutu.

Plants

The RAP botanical team made 401 plant collections representing 62 families, 132 genera, and approximately 240 species. These collections were made in the nine vegetation types we distinguished: tall herbaceous swamp vegetation and swamp wood, seasonally flooded forest, (seasonal) palm swamp forest, high tropical rainforest on dryland (terra firme), tropical forest on laterite/granite hills, savannah (moss) forest, open rock (inselberg) vegetation, secondary vegetation, and bamboo forest. We found eight species previously unrecorded in Suriname, of which six were tree species and two were herbaceous species. We also found a substantial number of rare plant species for Suriname, including six tree species listed on the IUCN Red List and three tree species protected under Surinamese law. The three sampling sites each had a distinct species composition, and the forests along the Kutari River had one of the highest tree alpha diversity values ever recorded for Suriname. At the same time, the forests at Werehpai had relatively low tree alpha diversity values. The forests showed some floristic affinities with adjoining regions of Guyana and Brasil. Comparison of our results with data from forests in northern Suriname showed that forests in the Kwamalasamutu region overlap only partially in species composition. Based on these results we argue that the forests in the Kwamalasamutu region have a high natural value for Suriname, and appropriate conservation measures should be taken, including the establishment of additional community-managed protected areas, and

exploration of agricultural methods that better incorporate standing forests.

Aquatic Beetles

We collected more than 4000 aquatic beetle specimens using both active and passive collecting techniques. We documented 144 species, representing 62 genera in nine families. Sixteen of these species have been confirmed as new, with an additional 10 likely to be new. Two of these new species, both in the family Hydrophilidae, are described here: *Oocyclus trio* Short & Kadosoe sp.n. and *Tobochares sipaliwini* Short & Kadosoe sp.n. Camps 1 (Kutari) and 3 (Werehpai) had comparatively high species diversity, with 91 and 93 species respectively—although only 48 of these species were shared between the two sites. Camp 2 (Sipaliwini) had the lowest number of species with 68. The fauna was typical of lowland Guianan forests. Some taxa, such as the genera *Siolus, Guyanobius, Fontidessus,* and *Globulosis* are either endemic or largely restricted to the Guiana Shield. The fauna was very similar to what is known from southern Venezuela (south of the Orinoco) and Guyana. The water beetle diversity was expected given the complement of aquatic habitats available at each camp. The relatively high number of genera and species, which cover a variety of ecological and habitat types, suggest the area is largely undisturbed.

Dung Beetles

Dung beetles are among the most cost-effective of all animal taxa for assessing biodiversity patterns, but relatively little is known about the dung beetle fauna of Suriname. I sampled dung beetles using baited pitfall traps and flight intercept traps in the Kwamalasamutu Region of southern Suriname. I collected 4,554 individuals represented by 94 species. Species composition and abundance varied quite strongly among sites. Dung beetle diversity correlated positively with large mammal species richness, and was highest at the most isolated site (Kutari), suggesting a possible cascading influence of hunting on dung beetles. Small-scale habitat disturbance also caused local dung beetle extinctions. The dung beetle fauna of the Kwamalasamutu region is very rich relative to other lowland forests of Suriname and the Guianas, and contains a mix of range restricted endemics, Guiana Shield endemics, and Amazonian species. I estimate that about 10–15% of the dung beetle species collected here are undescribed. While most species were coprophagous, 26 species were never attracted to dung; 4 of these were attracted exclusively to carrion or dead invertebrates and the other 22 were only captured in flight intercept traps. The abundance of several large-bodied dung beetle species in the region is indicative of the intact wilderness that remains. These species support healthy ecosystems through seed dispersal, parasite regulation and other processes. Maintaining continuous primary forest and regulating hunting (such as through hunting-restricted reserves) in the region will be essential for conserving dung beetle communities and the ecological processes they sustain.

Ants

At least 100 species of ants (Hymenoptera: Formicidae) were recorded around the Werehpai caves during the RAP survey. While ant data from the RAP survey are still being analyzed, a preliminary look at the ant fauna of the area indicates that the forests contain a diverse and abundant ant fauna. The presence of many dacetine species typical of closed-canopy rainforest indicate that the forests are in good condition. The ant fauna of Suriname is still very poorly known, with about 350 species documented, as few locations have been sampled for ants. Data on the ant fauna of the Kwamalasamutu area are valuable for eco-tourism since ants are easy to find and observe in the forest. Biological data for charismatic ant species will inform tourists about the hidden fauna of the rainforest and their important roles in ecosystem function and conservation. Habitat loss, fragmentation and the introduction of invasive ant species are the biggest threats to the ant fauna of the region.

Katydids

Seventy-eight species of katydids (Orthoptera: Tettigoniidae) were recorded during the RAP survey. At least seven species are new to science, and 29 species are recorded for the first time from Suriname, bringing the number of species of katydids known from this country up to 85. Of the three main camps, the Kutari site had the lowest number of both species (25) and specimens (64) collected, presumably because of the heavy rains that still affected the activity of katydids at the end of the rainy season, when the survey began. Werehpai had the highest number of species (54), followed by Sipaliwini (46). This RAP survey confirms that the katydid fauna of Suriname is exceptionally rich, yet still very poorly known. Although no specific conservation issues have been determined to affect the katydid fauna, habitat loss in Suriname due to logging and mining activities constitute the primary threat to the biota of this country.

Dragonflies and Damselflies

Ninety-four species of dragonflies and damselflies were recorded during the RAP survey, representing one-third of the species known from Suriname. Fifty-seven species were found at the Kutari River site, 52 at the Sipaliwini River site, and 65 at the Werehpai site. Fourteen species represent new records for Suriname, of which four, belonging to the genus *Argia*, are new to science; an additional five species represent first records at a new locality since their original descriptions, increasing considerably their known extent of occurrence. In terms of odonate community composition, the three sites shared between 1/2 and 2/3 of the species with each other, though relative abundance differed among sites. The diversity of odonate genera and species found in this study is characteristic of intact tropical lowland forest; most of the species found in the forest understory, creeks, and swamps would not be present if the forest were disturbed. Therefore it is recommended to designate a large and legally protected nature preserve to conserve the high diversity of odonate species found in this study. If forest cover and stream morphology are maintained in the area, the present odonate assemblages are expected to persist.

Fishes

We recorded 99 species of fishes from 43 sampling localities along the Sipaliwini and Kutari Rivers. This diversity is high compared to the rest of the world, but is typical for the Guiana Shield. We collected eight species of fishes potentially new to science, including a large catfish with spines along the body and a small catfish that lives in sand-bottomed creeks. Two species are new records for Suriname. We collected 57 species at the Kutari site, 60 species at the Sipaliwini site, and 63 species at the Werehpai site. This is remarkably consistent, with no significant difference in diversity among camps. However, we did not necessarily find the same species at each camp. Creek assemblages were similar among the three sites. Many young fishes were found in flooded forests, even if the adults lived in rivers or other habitats. Overall, large top-level predators were uncommon. The region is exhibiting the first stages of overfishing. Many fishes still occur in the Sipaliwini area, but there is a need to assess fishing pressure and implement management plans.

Reptiles and Amphibians

The RAP team found 42 species of amphibians and 36 species of reptiles, including a frog in the genus *Hypsiboas* that is new to science. The amphibian community was most similar to those of forests on bauxite plateaus in central Suriname. Several rare species were collected during the survey: *Osteocephalus cabrerai* is a rare tree frog from the western Amazon Basin and French Guiana and is reported from Suriname for the first time. *Scinax proboscideus* is a tree frog previously known from only two localities in the interior of Suriname and a few localities in French Guiana. *Microcaecilia taylori* was described, based on three specimens, from forest islands in the Sipaliwini savanna; the specimen collected by us is the fourth specimen known to science, and shows that this species is not restricted to the Sipaliwini savanna area. The snake *Xenodon werneri* is quite rare, and our record constitutes the third specimen for Suriname. The amphisbaenian *Amphisbaena slevini* was collected in Suriname for the first time. We also encountered *Chelonoides denticulata* (Yellow-footed Tortoise), listed as Vulnerable on the IUCN Red List. We discovered that certain expected species that are quite common in other areas in Suriname were either not found or found in very moderate numbers on the RAP survey. On the other hand, we found certain generally rare species to be quite common.

Birds

The RAP team recorded 327 species of birds: 294 species from the three RAP sites, 12 species observed in the area during the reconnaissance trip (3–8 May 2010) but not during the RAP survey, and 21 species observed only in the vicinity of Kwamalasamutu itself. The avifauna was typical of lowland forests of the Guiana Shield, and included many species endemic to the region. Our observations represent the first published records for Suriname of *Crypturellus brevirostris* (Rusty Tinamou), *Dromococcyx pavoninus* (Pavonine Cuckoo), *Xiphocolaptes promeropirhynchus* (Strong-billed Woodcreeper), and *Ramphotrigon megacephalum* (Large-headed Flatbill). The overall species list was highest for the Sipaliwini camp (250 species), followed by Werehpai (221 species) and Kutari (216 species). 153 species, or approximately 52% of those encountered at the three sites, were observed at all sites. The Kutari site had the most distinctive avifauna of the three sites. We estimate that a minimum of 350 bird species, or roughly half of the number of species known to occur in Suriname, may be found in the Kwamalasamutu area. Although no species listed on the IUCN Red List were encountered during the RAP survey, at least one (*Harpia harpyja*, Harpy Eagle, Near-Threatened) is known to occur in the area. Maintenance of large tracts of intact forest is recommended to preserve the avian diversity of the Kwamalasamutu region.

Small Mammals

The RAP team documented 38 species of small mammals including 26 species of bats, 10 species of rats, and two species of opossums. The species diversity and relative abundance of rats and mice at the three survey sites were the highest recorded in 20 years of mammal surveys throughout Suriname and Guyana by the Royal Ontario Museum. The Kutari site had the highest capture rate of rats and mice, indicating a healthy source of prey species for predators such as cats, owls, and snakes. In contrast, Werehpai was the most successful site for bats, but this was attributable to the well-established trails at the site, which functioned as flyways that were more conducive to capture success compared to the other sites, where rudimentary trails were only recently cut. This indicates that bats are relatively tolerant to minor alternations to their habitat. Noteworthy records include two species endemic to the Guiana Shield, a water rat (*Neusticomys oyapocki*) and a brush-tailed rat (*Isothrix sinammariensis*), collected at Kutari that represent the first occurrences of these species in Suriname. The primary conservation recommendations arising from the small mammal survey of the Kwamalasamutu region are: 1) designation of the Kutari area as a nature reserve because of the high species diversity and relative abundance of rats and mice that are necessary to sustain healthy populations of top-level predators; and 2) minimal development of the Werehpai petroglyph site to ensure continued ecosystem services of the bat fauna including seed dispersal, flower pollination, and insect control.

Large Mammals

Twenty-nine species of medium- and large-bodied mammals were recorded through visual encounters and camera trapping. Large caviomorph rodents were the most frequently recorded animals in the camera trap images. The Kutari site was the richest in species, especially primates. The Brazilian

Tapir (*Tapirus terrestris*, IUCN Vulnerable) was recorded by the camera traps at all three sites and was observed by several of the RAP scientists. Of the six species of cats known to occur in the Guiana Shield region, the Jaguar (*Panthera onca*, IUCN Near-Threatened), Puma (*Puma concolor*) and Ocelot (*Leopardus pardalis*) were found during the survey. The White-lipped Peccary (*Tayassu pecari*, IUCN Near-Threatened) was only photographed once by the camera traps in the Werehpai area and seems to be uncommon in the Kwamalasamutu region. In addition to the species mentioned above, four additional species listed on the IUCN Red List were encountered: *Ateles paniscus* (Guianan Spider Monkey, Vulnerable); *Myrmecophaga tridactyla* (Giant Anteater, Vulnerable); *Priodontes maximus* (Giant Armadillo, Vulnerable); and *Pteronura brasiliensis* (Giant Otter, Endangered). The number of mammal species found during this survey does not differ much from what was expected. The difference in number of species per site suggests that hunting pressure varies from one area to another. Our observations of many shy and sensitive mammal species indicate that hunting has not yet depleted game populations in the region. Nevertheless, hunting from Kwamalasamutu represents the most significant current threat to medium- and large-bodied mammals in the area. Recommended studies include more camera trapping and a sustainability evaluation of wild meat hunting.

SUMMARY OF CONSERVATION RECOMMENDATIONS

Conservation Action

Establish protected areas to maintain the intact ecological condition of the area's forests and rivers. Monitor and prevent illegal mining activity in the Kwamalasamutu region.

The results of the RAP survey indicate that the Kwamalasamutu region is in near-pristine ecological condition. The area supports high species diversity, including many species found only in extensive regions of undisturbed forest. We found no evidence of substantial anthropogenic impacts on water quality or forest structure away from the village itself. As the forested landscape of this area extends unbroken far beyond the borders of Suriname, the Kwamalasamutu region represents the nucleus of a vast biological treasure of global significance. Although not immediately threatened, effective conservation in the region will require active and continuous assessment of potential threats and international cooperation to adequately manage the region's resources.

We attribute much of the region's high species diversity to small-scale habitat heterogeneity and intact connections between habitats used by animals in different stages of their life cycles. Even within primary terra firme forest, most taxonomic groups showed surprisingly high species turnover between sites. This mosaic of diversity is typical of large, undisturbed regions of tropical forest, and can be impacted profoundly by human modification of the landscape. At a large spatial scale, road construction and resource extraction (e.g., logging, mining) should be carefully controlled to avoid disrupting processes vital to maintenance of ecosystem integrity. At a smaller scale, guidelines should be developed for establishing protected areas that consider fine-scale environmental heterogeneity as well as the seasonal movement of animals among different habitats, particularly aquatic and terrestrial habitats.

Of particular concern is the continuing encroachment of small-scale gold miners in the region, which can be expected to accelerate with the construction of highways currently planned for interior Suriname and adjacent northern Brazil. The clean and abundant water flowing from the upper Corantijn watershed is an extremely valuable asset, both for the people who depend directly on the rivers for sustenance and for the people of coastal Suriname. Pollution of rivers by small-scale miners, a persistent problem elsewhere in the Guianas, has the potential to cause major ecological and social upheaval in the Kwamalasamutu area if miners gain access to the region. Already there are concerns among residents of Kwamalasamutu about gold mining activities in the upper reaches of the Aramatau River, and our data suggest that mercury pollution may already be affecting the region's watercourses (see Chapter 1, Water Quality). Aside from mercury contamination, any increase in mining activity would contribute to erosion and sedimentation, negatively impacting fish stocks upon which the people of the region depend. In addition, gold miners often hunt intensively and can severely deplete populations of terrestrial mammals that indigenous people, as well as healthy ecosystems, depend upon (see below).

Environmental Protection and Sustainable Harvesting
Develop and implement a plan to manage bush meat hunting and fishing in the Kwamalasamutu region.
Effective conservation in the Kwamalasamutu region will require active management of wildlife and their habitats to protect them from overexploitation. This is particularly important if the community desires to pursue ecotourism as a source of revenue (see below). Already there are signs that wildlife has been impacted, especially near the village. Our strongest evidence for this is the observation that large, predatory fishes, many of which are prized for food (e.g., *Hoplias aimara*, *Cichla ocellaris*), were generally scarce even at the most remote camp, and virtually absent in the vicinity of Kwamalasamutu. Although the camera traps and dung beetle surveys respectively provided direct and indirect evidence for a rich mammal fauna, the general scarcity and shyness of wildlife (particularly monkeys, peccaries, and curassows) at all sites was suggestive of hunting pressure. High mammal and dung beetle diversity at the most isolated site (Kutari) also suggests that hunting closer to the village is impacting local ecosystems. Populations of game animals in the region are probably sustained by dispersal through the vast and largely uninhabited forest matrix that surrounds the Kwamalasamutu region, where we presume wildlife is more abundant. However, this does not justify local depletion of

wildlife, as many game animals and fishes play important roles as predators and seed dispersers in the ecosystem, and as such are vitally important for forest dynamics.

We suggest that a thorough assessment of bush meat hunting and fishing pressure be undertaken to promote establishment of, and adherence to, hunting and fishing quotas or seasons for particular species. Ideally, this would incorporate information on the ecology and reproductive habits of target species, already well known to many residents of the region. Alternatively, certain areas could be designated as non-hunting zones for at least a portion of each year, following the model of the Iwana Samu sanctuary set up by the people of Kwamalasamutu. However, the effectiveness of these protected areas depends on diligent local enforcement of activities within them. By either of these mechanisms, the regulation of bush meat hunting would benefit the residents of Kwamalasamutu by allowing wildlife populations to replenish themselves, thereby lessening the need for expensive hunting excursions far from the village. Chickens and other domestic animals also provide a good alternative protein source, and should be further encouraged in the village.

Ecotourism: Promotion & Implementation
Continue developing and upgrading the Iwana Samu ecotourism facilities, focusing on the region's cultural history.
Ecotourism has great potential to provide the village of Kwamalasamutu with much-needed income. To this end, the community should enhance the existing facilities at Iwana Samu and work to highlight the uniqueness of the area, manifested in the petroglyphs at Werehpai and elsewhere. Protection of wildlife (see above) would also help increase the area's appeal to tourists, many of whom will require some incentive to choose to visit Kwamalasamutu in lieu of less expensive destinations closer to Paramaribo. Protection of fish stocks could allow the development of sport fishing tourism. Adventure tourism (e.g. trekking) could also be promoted by taking advantage of the existing network of trails used by residents of the region to move between settlements. Effective advertisement and promotion of the site and facilities to tourists in the Netherlands, United States and other countries will also be key to the success of ecotourism here.

The data from this RAP survey are being incorporated into an educational/tourism booklet about the biodiversity of the Kwamalasamutu region that can be used by the people of Kwamalasamutu in their eco-tourism efforts. Charismatic species of birds, mammals, amphibians, dragonflies and other taxa — even ants! — have been identified and will be promoted as key attractions for tourists.

Scientific Capacity Building

Develop research facilities to promote the exchange of information between residents of Kwamalasamutu and scientists from Suriname and abroad. Develop and implement a water quality monitoring protocol.

The Kwamalasamutu community would benefit from the development of facilities for Surinamese and foreign researchers. The region supports a high diversity of aquatic and terrestrial habitats and is relatively free of large-scale anthropogenic degradation, rendering it highly suitable for ecological research. We consider the area to be particularly promising for research on the ecological role of humans in tropical lowland forest, given the region's long history of occupation by the Trio. To this end, researchers could employ and train residents of Kwamalasamutu in a mutually beneficial relationship, whereby researchers gain valuable field assistance and indigenous knowledge in exchange for site-specific recommendations for management of natural resources to promote long-term social and environmental stability. In particular, we recommend that residents of Kwamalasamutu be trained to implement a water quality monitoring program to empower them to detect and act upon the first signs of degradation of this vital resource. Residents could also be monitoring and recording wildlife observations and bushmeat and fish consumption patterns in order to gain better understanding of the long-term dynamics and sustainability of hunting and fishing.

One of the greatest potential threats to the region is the erosion of traditional knowledge among young people. We recommend creating educational materials—for example, picture guides to common species of birds, fishes, and mammals—to be translated into Trio and used in area schools. These guides could also be used by tourists visiting the region.

Further Studies

Conduct additional biodiversity surveys at different times of the year.

Although we found a high diversity of species in the Kwamalasamutu region, our survey was only the first step toward a thorough knowledge of the region's biodiversity. Beyond documenting species new to science, biodiversity surveys provide critical baseline information about the distribution, ecology, and habitat requirements of tropical organisms. Many tropical plants and animals are poorly known from a scientific perspective; this is particularly true for the species new to science that we encountered on this survey. We therefore recommend additional surveys, focusing on under-sampled habitats (e.g. inselbergs), different seasons, and other sites within the region, to gain a better understanding of the biodiversity of the Kwamalasamutu region and southwest Suriname in general. We suspect that many undescribed species remain to be discovered.

REFERENCES

Alonso, L.E. and J.H. Mol (eds.). 2007. A rapid biological assessment of the Lely and Nassau plateaus, Suriname (with additional information on the Brownsberg Plateau). RAP Bulletin of Biological Assessment 43. Conservation International, Arlington, VA, USA.

Alonso, L.E., J. McCullough, P. Naskrecki, E. Alexander, and H.E. Wright (eds.). 2008. A rapid biological assessment of the Konashen Community Conservation Area, Southern Guyana. RAP Bulletin of Biological Assessment 51. Conservation International, Arlington, VA, USA.

Alonso, L.E. and H.J. Berrenstein (eds.). 2006. A rapid biological assessment of the aquatic ecosystems of the Coppename River Basin, Suriname. RAP Bulletin of Biological Assessment 39. Conservation International, Washington, DC.

Bernard, E. (ed.). 2008. Inventários Biológicos Rápidos no Parque Nacional Montanhas do Tumucumaque, Amapá, Brasil. RAP Bulletin of Biological Assessment 48. Conservation International, Arlington, VA, USA.

Hammond, D. S., ed. 2005. Tropical Forests of the Guiana Shield: Ancient Forests in a Modern World. Oxfordshire, UK: CABI International.

Hollowell, T., and R. P. Reynolds. 2005. Checklist of the terrestrial vertebrates of the Guiana Shield. *Bulletin of the Biological Society of Washington* 13.

Huber, O., and M. N. Foster. 2003. Conservation Priorities for the Guayana Shield: 2002 Consensus. Washington, DC: Conservation International.

IUCN 2011. IUCN Red List of Threatened Species. Version 2011.1. www.iucnredlist.org.

Sandoval, A.E. 2005. Preliminary report on ancient human occupations at Werehpai, Southern Suriname. Report to Conservation International-Suriname.

Teunissen, P., and D. Noordam. 2003. Ethno-ecological survey of the lands inhabited/used by the Trio people of Suriname, Part 1: Ecological survey. Arlington, VA: Amazon Conservation Team.

Vari, R.P., C.J. Ferraris, Jr., A. Radosavljevic, and V.A. Funk. 2009. Checklist of freshwater fishes of the Guiana Shield. *Bulletin of the Biological Society of Washington* 17.

Chapter 1

A baseline water quality assessment of the Kutari and Sipaliwini Rivers

Gwendolyn Landburg and Mercedes Hardjoprajitno

SUMMARY

We sampled water quality at 23 sites on the Kutari, Sipaliwini, and Aramatau Rivers. The oxygen content and pH of the Kutari River were lower than those of the Sipaliwini River, probably due to the lack of rapids and the input of organic material from the surrounding forest, particularly after heavy rains, which occurred frequently at the Kutari site. All sites had clear water except the Wioemi Creek, which was very turbid. The parameters measured in the field revealed undisturbed river ecosystems with few negative human impacts. However, high mercury levels were found in both sediment and piscivorous fishes from all sites. Further research is needed to clarify the origin of mercury in the rivers of southwest Suriname. Suggestions are given for a water quality monitoring program that can be implemented by the residents of Kwamalasamutu.

INTRODUCTION

Water is important for all living creatures. The type and quality of water determines which organisms will be found in certain habitats. Assessment of water quality is needed to identify species-habitat relationships and possible sources of pollution or disturbance within the ecosystem. For this reason, it is necessary to gather baseline data on basic environmental parameters (pH, dissolved oxygen, conductivity, turbidity) as well as levels of nutrients (as indicators of nutrient cycling in the surrounding ecosystem) and metals (as indicators of pollution or erosion from underlying bedrock).

Human disturbance was not expected in the area assessed by the Kwamalasamutu RAP survey, but previous studies have discovered mercury pollution in otherwise pristine areas of Suriname. It has been hypothesized that mercury might be transported by the northeast trade winds from gold mining sites in eastern Suriname to the southwestern region of the country (Landburg 2005, P.E. Ouboter *unpubl. data*), or, alternatively, that mountain ranges in central Suriname serve as a barrier, resulting in mercury deposition on the windward side of the mountain ranges and no deposition on the leeward side. A primary goal of this study was to provide baseline information on mercury levels in southwest Suriname to further evaluate these hypotheses.

STUDY SITES AND METHODS

Twenty-three sites were sampled intensively in three major areas: the Kutari River, and two areas of the Sipaliwini River (Fig. 1). The Kutari River can be characterized as a clear water river without major rapids at the time of sampling. The river extends into the forest when the water level increases, resulting in major floodplains in the area. Big creeks flowing into the Kutari River have steep banks and smaller floodplains. The two sampled areas of the Sipaliwini

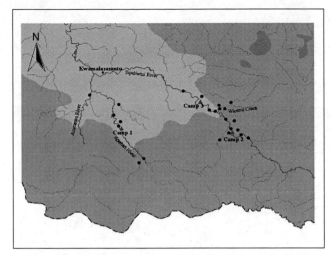

Figure 1. Sites sampled in the Kutari River and two areas of the Sipaliwini River.

River share many characteristics including steep riverbanks, clear water, and much turbulence in the water, caused by the many rapids in the river. The Wioemi Creek is a large creek with especially turbid water, fairly steep banks, a strong current, and moderately extensive floodplains. Other creeks flowing into the Sipaliwini River are clear water streams, with steep banks and weak currents. We also sampled one site near the mouth of the Aramatau River, which was similar to the Kutari River but had steeper banks.

We measured 13 physico-chemical parameters at each site: pH, dissolved oxygen, conductivity, temperature, alkalinity, total hardness, total phosphate, nitrate, chloride, tannin & lignin, ammonia, turbidity, and secci depth (Appendix A). Both titrimetric and colorimetric methods were used to assess the parameters. At selected sites, water samples were saved for later analysis of mercury, iron, and aluminum at the University of Suriname in Paramaribo (Appendix B). For mercury analyses, sediment and fish tissue samples were taken opportunistically. All stored samples were kept under refrigeration in the field.

RESULTS

Summary data on baseline parameters for each site are presented in Appendix A. Measurements of nutrients, salts, and metals from each site are presented in Appendix B.

Kutari and Aramatau Rivers. The oxygen content in the Kutari River and tributary creeks was lower than the other sites (4.2–5.4 mg/L), probably a result of the lack of rapids and the input of organic material from the land. The pH was lower at these sites as well (range 5.6–5.9). Nutrient input comes mainly from the land, as evidenced by the higher nutrient levels measured in the water after heavy rain (phosphate: 0.03–0.1 mg/L; ammonia: 0.26–0.72 mg/L). High levels of mercury were found in both sediment

(0.26–0.28 µg/g) and piscivorous fishes (0.05–0.98 µg /g). These levels are higher than the Canadian Interim Sediment Quality Guideline for Protection of Aquatic Life of 0.17µg/g soil, and the European Union standard for human consumption of 0.5 µg/g fish (Canadian Council of Ministers of the Environment 1999; EC 2002).

From the one site sampled in the Aramatau River, low levels of nutrients were measured (phosphate 0.04 mg/L; nitrate: 0.00 mg/L) except for ammonia (average: 0.43 mg/L). The water at this site was found to be very soft (hardness: 0.35 mg/L). High mercury levels were found in the sediment (average: 0.19 µg/g).

Sipaliwini River. At the sites in the Sipaliwini river and tributary creeks, high nutrient levels were measured (phosphate: 0.045–0.145 mg/L; ammonia: 0.51 mg/L average). We also found high levels of iron (0.98–1.29 mg/L). Mercury levels were low in water (0.03–0.07 µg/L) compared to the EPA drinking water standard of 2 µg/L (US EPA 1994), whereas sediment and fish contained higher mercury levels (sediment: 0.12–0.19 µg/g; fish: 0.28–1.17 µg/g).

Wioemi Creek. The strong current and consequent erosion of the steep banks at the time of sampling probably contributed to high turbidity (average turbidity: 22.08 NTU) and nutrient loads (average nitrate: 0.013 mg/L; average phosphorus: 0.105 mg/L; average ammonium: 0.85 mg/L), as well as high levels of aluminum (0.89–1.12 mg/L) and iron (1.56–1.73 mg/L). Mercury levels in the water were low (0.00–0.03 µg/L), while mercury levels in sediment were high (0.18–0.25 µg/g).

DISCUSSION AND CONCLUSIONS

In general, our data revealed river ecosystems with relatively clear water (except Wioemi Creek), high nutrient loads from the surrounding flooded forests, and high levels of metals. High levels of iron and aluminum are usually attributed to natural erosion of the bedrock or anthropogenic activities. Because the area sampled is largely free of large-scale anthropogenic disturbance, the levels of these metals are probably a natural consequence of eroded bedrock material entering the aquatic system.

The high mercury levels found in the ecosystem suggest that small-scale gold mining in eastern Suriname is affecting this area. We know of no gold mining activities in the Kwamalasamutu region, though some residents of Kwamalasamutu expressed concern about gold mining upstream of the Aramatau River. This needs further investigation. The high mercury levels measured in the other rivers indicate that mercury is probably transported through the atmosphere and is being deposited in these systems. Further research is needed to confirm this. Deposition of mercury in these areas may result in accumulation of mercury in the food chain, causing health concerns for residents of Kwamalasamutu.

RECOMMENDATIONS

Because the assessed area is very important for the residents of Kwamalasamutu, it is essential to measure some water quality parameters (as indicators) regularly. Parameters selected need to indicate pollution sources and the safety of drinking water. The monitoring may start with regular measurements of pH, dissolved oxygen, conductivity, temperature, turbidity and/or secci depth, and bacteria (*Escherichia coli*). These measurements should be done every three months. More extensive sampling should be done twice per year—once in the dry season (September–November) and once in the wet season (May–August)—and should include assessments of mercury in river sediments and fishes.

REFERENCES

Canadian Council of Ministers of the Environment. 1999. Canadian sediment quality guidelines for the protection of aquatic life: Mercury. In: Canadian Environmental Quality Guidelines, 1999, Canadian Council of Ministers of the Environment, Winnipeg.

EC. 2002. EC Regulation (221/2002) amending Commission Regulation (EC) no. 466/2001 of 8 March 2001 setting maximum levels for certain contaminants in foodstuffs.

Landburg, G. 2005. Kwikvervuiling in midden en west Suriname; natuurlijk of antropogeen (Mercury pollution in central and west Suriname; Natural or anthropogenic). B.Sc. Thesis, Anton de Kom University of Suriname.

US EPA. 1994. Water Quality Standards Handbook. Second Ed. USEPA Water Resource Center. United States Environmental Protection Agency EPA-823-B-94-005. EPA, Washington, DC.

Appendix A. Water Quality Data – Basic Parameters. Legend: Cond = conductivity; DO = dissolved oxygen; Secci = maximum depth at which markings on Secci disc could be read.

Location	Location name	Cond (µS/cm)	pH	DO (mg/L)	DO (%)	Alkalinity (mg/L CaCO$_3$)	Hardness (mg/L CaCO$_3$)	Tannin Lignin (mg/L)	Turbidity (NTU)	Secci (cm)
01-01	Big creek downstream Koetari river	14.4	6.415	6.25		7.25		1.3	13.25	current too strong
01-02	Koetari river downstream camp 1a	11.4	5.625	5.15		4.85		1.1	9.425	current too strong
01-03	Creek downstream camp 1	10.6	5.785	4.5		3.4	2.85	1	10.01	56.75
01-04	Creek upstream camp 1	11.2	5.805	4.2		6.45	2.35	1.05	6.835	83.75
01-05	Koetari river upstream camp 1	11.4	5.725	5.2		5.6	1.25	1.05	7.39	current too strong
01-06	Koetari river downstream camp 1	11.9	5.87	5.4		5.5	1.3	1.15	8.82	current too strong
01-07	Aramatau river downstream	12.2	6.23	6.5		5.7	0.35	1.05	4.615	1.2
02-01	Sipaliwini river Upstream camp 2	20.6	6.775	7.1	90	8.1	1.95	1.05	0.535	0.535
02-02	Creek upstream camp 2	16.9	6.58	6.85	84.5	7.35	1.85	0.65	1.055	110
02-03	Sipaliwini river upstream camp 2	21	6.655	7.2	92	9.3	2	1.15	4.275	current too strong
02-04	Sipaliwini river downstream camp 2a	21.3	6.545	7.7	98.5	8.9	2.6	0.95	1.15	current too strong
02-05	Creek downstream camp 2 (a)	15.45	5.725	6.3	76	7.05	1.65	1	3.75	77.5
02-06	Creek downstream camp 2 (b)	18.2	6.175	6.75	82.5	8.35	2.4	1.4	0.01	0.01
02-07	Sipaliwini river downstream camp 2	21.4		7.1	91	8.55	2.35	0.85	4.99	current too strong
02-08	creek downstream camp 2	20		6.6	82	9.75	2.25	0.85	0	90
03-01	Sipaliwini river upstream camp 3 (a)	21.4	6.905	7.1	90	8.9	2.8	0.85	5.21	current too strong
03-02	Sipaliwini river-upstream camp 3	22.2	6.42	6.85	87	8.45	2.65	1.25	5.79	66.25
03-03	Creek downstream camp 3	20	6.31	6.6	80	8.85	4.6	1.3	6.695	66.25
03-04	Sipaliwini river downstream camp 3	21.8	6.79	6.7	85	10.4	3.4	0.95	10.9	current too strong
03-05	Creek Wioemi midstream	21.7	5.695	2.7	33	8.75	2.25	0.85	8.855	67.5
03-06	Creek Wioemi upstream	22.9	6.29	6.35	78.5	10.65	2.4	1.1	20.55	40
03-07	Creek Wioemi downstream (a)	22.7	6.2	5.9	73	10.4	2.45	1.15	23.6	45
03-08	Creek Wioemi downstream (b)	16.5	5.76	5.6	72	9.55	2.2	0.85		97.5

Appendix B. Water Quality Data – Nutrients, Salts and Metals.

Location	Location name	PO$_4$ (mg/L)	NO$_3$ (mg/L)	Ammonia (mg/L)	Chloride (mg/L)	Aluminium (mg/L)	Iron (mg/L)	Hg in water (mg/L)	Hg in sediment (µg/g)	Hg in fish (µg/g) Average per site
01-01	Big creek downstream Koetari river	0.1	0.01	1	2.5		0.37	0.03		
01-02	Koetari river downstream camp 1a	0.045	0.035	0.72	2.85		1.17	0.03	0.28	
01-03	Creek downstream camp 1	0.06	0.045	0.415	3.7					
01-04	Creek upstream camp 1	0.015	0.02	0.26	4.8					
01-05	Koetari river upstream camp 1	0.03	0.01	0.315	4.6		1.09	0.02	0.26	0.58
01-06	Koetari river downstream camp 1	0.075	0.005	0.7	3.65			0.03		
01-07	Aramatau river downstream	0.04	0	0.425	4.65		0.48	0.03	0.19	
02-01	Sipaliwini river Upstream camp 2	0.125	0.01	0.535	6		1.17	0.05	0.15	
02-02	Creek upstream camp 2	0.085	0.01	1.055	8.5	0.71	1.02	0.03	0.14	0.80
02-03	Sipaliwini river upstream camp 2	0.08	0.01	0.585	6.55					0.54
02-04	Sipaliwini river downstream camp 2a	0.065	0.025	1.15	7.85	0.51	1.24	0.07	0.16	0.76
02-05	Creek downstream camp 2 (a)	0.045	0	0.395	7.6	0.57	0.98	0.06	0.23	
02-06	Creek downstream camp 2 (b)	0.08	0.01	0.375	7.9				0.26	
02-07	Sipaliwini river downstream camp 2	0.11	0.02	0.21	7.8				0.12	
02-08	creek downstream camp 2	0.065	0	0.765	6.55	0.59	0.99	0.03		
03-01	Sipaliwini river upstream camp 3 (a)	0.09	0.01	0.27	8.2			0.07		
03-02	Sipaliwini river- upstream camp 3	0.12	0.015	0.755	9.65			0.07		
03-03	Creek downstream camp 3	0.055	0.01	0.285	7.45	0.90	1.28	0.06	0.27	0.42
03-04	Sipaliwini river downstream camp 3	0.145	0.005	0.705	8.35	0.79	1.29	0.08	0.15	
03-05	Creek Wioemi midstream	0.08	0	0.45	7.6	0.89	1.15	0.03	0.26	
03-06	Creek Wioemi upstream	0.145	0	1.04	7.3	1.02	1.56		0.24	
03-07	Creek Wioemi downstream (a)	0.065	0.025	0.66	8.8	1.12	1.73	0.02	0.22	
03-08	Creek Wioemi downstream (b)	0.065	0.015	0.535	6.95	0.73	1.20	0.00	0.22	

Chapter 2

Plant diversity and composition of the forests in the surroundings of Kwamalasamutu

Olaf Bánki and Chequita Bhikhi

SUMMARY

During a rapid assessment of the plant diversity and composition of the forests in the surroundings of Kwamalasamutu we made 401 plant collections belonging to 62 families, 132 genera, and approximately 240 species. These collections were made in the nine vegetation types we distinguished. We found eight species previously unrecorded in Suriname, of which six were tree species, and two were herbaceous species. We also found a substantial number of rare plant species for Suriname, including six tree species listed on the IUCN Red List and three tree species protected under Surinamese law. The forests in the surroundings of Kwamalasamutu are heterogeneous, and different forest types can be found in close proximity to one another. The forests at the three sampling sites each had a distinct species composition. The forests along the Kutari River had one of the highest tree alpha diversity values ever recorded for Suriname. At the same time, the forests at Werehpai had relatively low tree alpha diversity values. Comparison of our results with data from forests in northern Suriname showed that forests in the Kwamalasamutu surroundings have to some extent a distinct species composition. Based on these results we argue that the forests in the surroundings of Kwamalasamutu have a high natural value, one that warrants appropriate conservation measures.

INTRODUCTION

Most of our knowledge of the diversity and species composition of different forest types is based on studies from the northern regions of Suriname, whereas the forests in the southern regions of Suriname (as well as Guyana and French Guiana) are relatively unexplored. We know very little about the diversity and species composition of these forests. This lack of knowledge of plant diversity and composition hampers the assessment of the natural value of forests in southern Suriname. This knowledge is very much needed for sound decision-making concerning the sustainable management of the forest in Suriname.

Very often the tropical lowland forests of the central and southern regions of the Guianas are considered as one uniform forest type. However, an analysis of 156 1-ha plots across the Guianas has demonstrated that tree diversity and composition of forests mostly follow geological formations (Bánki 2010). The northern regions of Suriname are mostly made up of new and old coastal plains with their specific vegetation and plant species composition (Lindeman and Moolenaar 1959). The Zanderij or Coesewijne formation, with its white and brown sands, separates the northern regions from the Guiana Shield basement complex that extends over the central and southern parts of Suriname. This change in geological formation is also (partly) reflected in a change in species composition and diversity, especially concerning the white sands (Bánki 2010). Extrapolations suggest that tree alpha diversity could be higher in southern Suriname compared to the northern regions of Suriname (ter Steege et al. 2006). Although results from niche modeling based on herbarium collections do suggest similar patterns, the findings also suggest that forests in the Kwamalasamutu region could be less diverse

(Haripersaud 2009). We are in need of quantitative data on tree diversity and species composition of the forests in southern Suriname.

Our objective during the rapid assessment in the surroundings of Kwamalasamutu was to provide baseline data on abundance and diversity of trees, and plant species in general, by vegetation surveys and plot inventories. These data are a first step in determining to what extent the diversity and species composition of the forest in the Kwamalasamutu surroundings differs from the northern region of Suriname. As the RAP sites are close to both Guyana and Brasil, we expected to encounter some new plant species for Suriname. For comparison we used extensive datasets on 1-ha plots (Bánki 2010) and 0.1-ha plots (O. Bánki unpublished data; C. Bhikhi unpublished data) assembled during previous years in the northern region of Suriname.

FLORISTIC TEAM

The floristic team consisted of the following people:
Dr. Olaf Bánki and Chequita Bhikhi
General Plant collecting, plot inventories, and species identification
Klassie Etienne Foon
Tree spotter, plot inventories
Aritakosé Asheja and Sheinh A Oedeppe
Local tree spotter, trainee, tree climbers
Reshma Jankipersad, Jonathang Sapa
Field assistants and trainee, plot inventories
Tedde Shikoei, Willem Joeheo, Mopi and Dennis
Boatmen, field assistants

METHODS

We sampled at three sites: the Kutari River (Site 1), the Sipaliwini River (Site 2), and at Werehpai and Wioemi Creek (Site 3). At each site we used the same sampling methods, consisting of general vegetation surveys and plot inventories.

General Vegetation Surveys
General plant collecting took place along trails in the forest and along the rivers, including while traveling between camps. All flowering and fruiting plants encountered were collected during these surveys, and we recorded the different vegetation types found. Plant collections were numbered in the number series of Olaf Bánki (using OSB), and were pressed and dried in the field above kerosene stoves. At least one duplicate was stored in the National Herbarium of Suriname. The other duplicates were sent to the National Center for Biodiversity Naturalis (NCB), which harbors all plant collections of the National Herbarium of the Netherlands. Identifications of the dried specimens took place in the Guianas collections of the NCB Naturalis. The specimens were identified using several plant identification books of the Guianas and plant identification keys (e.g., Pulle et al. 1932; Jansen-Jacobs 1985; Steyermark et al. 1995). After identification, the specimens were compared with herbarium specimens for confirmation. Duplicates of some plant families were sent to their respective taxonomic group specialist within the Flora of the Guianas network. We determined new records for Suriname by checking the occurrence of the species in the checklist of the Guianas (Funk et al. 2007) and the digital database of the Guianas collections of the NCB Naturalis, and by consulting the collections of the Missouri Botanical Gardens at www.discoverlife.org.

Plot Inventories
At each study site we created one 1-ha plot (250 × 40 m) in a dominant high forest type on dryland (terra firme) and identified all trees above 10 cm dbh (diameter at breast height) in the plots. Within each site, we placed one 0.1-ha plot several hundred meters from the 1-ha plot, in the same high forest type, and identified all tree species above 2.5 cm dbh. Palms were included in the assessment, whereas lianas were not assessed in these plots due to time constraints. Preliminary identification of trees in the plots was made by tree spotter Klassie Etienne Foon of SBB, Olaf Bánki, and ACT personnel Sheinh A Oedeppe and Aritakosé Asheja. For each species encountered for the first time in the plots, we made a plant collection. Collections of tree species were processed in a similar way as the plant collections of the general surveys.

In total, six plots were established (Table 1). The Kutari plots (Ku1 & Ku2) were established in high mature tropical rainforest on loamy sands. Soils were deep and well drained, and there were no boulders or traces of hard parent rock in the plot. The Sipaliwini plots (Si3 & Si4) were placed on a

Table 1. Metadata for the plots established at each site during the RAP. N = number of individuals, S = number of species, Fα = Fisher's alpha.

Plot Name	Ha	Dimensions	N	S	Fα	Lat	Long
Kutari River Plot 1	1	250 × 40 m	529	140	62.15	240377	524499
Kutari River Plot 2	0.1	100 × 10 m	142	81	78.3	240461	523639
Sipaliwini River Plot 3	1	250 × 40 m	443	116	51.14	252395	544182
Sipaliwini River Plot 4	0.1	100 × 10 m	123	54	36.74	253056	544157
Werehpai Plot 5	1	250 × 40 m	454	104	42.19	262877	535847
Werehpai Plot 6	0.1	100 × 10 m	158	46	21.8	262640	535547

hill approximately 200–300 meters above sea level. The forest was standing on shallow to deep loamy sandy soils on top of the hard parent rock of the Guiana Shield basement complex. A portion of plot Si3 contained forest transitioning into liana forests and low savannah forest due to large boulders and the hard parent rock reaching the surface. The Werehpai plots (We5 & We6) were placed in mature tropical rain forest on sandy soils with large boulders throughout the plots. Soils were deep at some points, but predominantly shallow on the hilltops because of the hard parent rock underneath.

Plot Comparison

To investigate the floristic and diversity differences between the forests in northern/central Suriname and the forests in the Kwamalasamutu surroundings, we compared the three 0.1-ha plots with unpublished 0.1-ha plot data from Olaf Bánki (6 plots from Gros Rosebel), Chequita Bhikhi (3 plots from the van Blommestein Lake, Brokopondo District), and Pieter Teunissen (12 plots from Gros Rosebel). We compared our data from the three 1-ha plots with data from 28 1-ha plots on the brown sands of the Zanderij/Coesewijne formation, and from the lowlands, slopes, and plateaus of the Brownsberg, the Lely Mountains, and the Nassau Mountains (Bánki 2010).

Plot analyses

The 1-ha and 0.1-ha plot datasets were analyzed as two separate datasets. Differences in floristic composition between plots were investigated with the ordination technique of Non-Metric Multi-Dimensional Scaling (NMS) with Relative Sörenson as the floristic distance measure, 250 real and randomized data runs, and 4–6 dimensions (NMS in PCORD 5; McCune & Grace 2002; McCune & Mefford 1999). We also performed a Detrended Correspondence Analysis on both the 0.1-ha and the 1-ha plot datasets (DCA in PCORD 5, McCune & Grace 2002; McCune & Mefford 1999). Only the results of the NMS are presented in this report. We also ran species indicator analyses on the 0.1-ha and 1-ha plot datasets to investigate which species were responsible for the division of the plots in several floristic groups (in PCORD 5; Dufrene & Legendre 1997; McCune & Grace 2002; McCune & Mefford 1999). The tree alpha diversity of the plots was expressed as Fisher's alpha (Fisher et al. 1943). Fisher's alpha is a diversity index describing the relation between the number of individuals and species in a plot. Differences in the averages of the number of species, number of individuals, and in Fisher's alpha were statistically tested through ANOVA (SPSS-Inc. 2007).

RESULTS

Vegetation Descriptions

At the three locations and between the camps, we encountered several vegetation types. Based on the vegetation descriptions of Lindeman and Moolenaar (1959) and Bánki

(2010), we distinguished a total of nine different vegetation types:

1. **Tall herbaceous swamp vegetation and swamp wood.** This vegetation type was abundant in the bends of rivers and creeks, and was found around all three study sites. The herb layer consisted mostly of dense stands of *Montrichardia arborescens* (mokumoku, Araceae) intertwined with cyper grasses, grasses, and vines. Most of the shrub and tree layer consisted of *Inga* sp. (watra switibonki, Fabaceae). Dense stands of *Inga* trees occurred in the river bends, as well as along the river edges. Solitary and clumped palm trees with spiny trunks (*Bactris* sp., Arecaceae), solitary trees of *Cordia* sp. (tafrabon, Boraginaceae) with table-like crowns, and solitary trees of *Cecropia* sp. (bospapaja, Cecropiaceae) occurred in swampy areas in the river bends. At Wioemi Creek and the Sipaliwini River site, we observed individual *Triplaris surinamensis* (mira udu, Polygonaceae) trees. At the Kutari River we observed one *Erythrina fusca* (kofimama, Fabaceae) tree in the swamp wood. This vegetation type as well as its species composition shows resemblance to the coastal areas in northern Suriname (see Lindeman & Moolenaar 1959).

2. **Seasonally flooded forest.** We observed seasonally flooded forests with quite different species composition. At the margins of the black waters of the Wioemi Creek and the Kutari River we observed forests dominated by *Tachigali paniculata* (mira udu, Fabaceae), *Alexa wachenheimii* (neku or paku nyannyan, Fabaceae), *Eperua rubiginosa* (oeverwalaba, Fabaceae), and different species of Myrtaceae, Sapindaceae, Meliaceae, and Annonaceae. Large areas of seasonally inundated forest were found along both the Wioemi Creek and the Kutari River. Along the Wioemi Creek, the seasonally flooded forest was dominated by *Astrocaryum sciophilum* (bugru maka, Arecaceae), *Licania* sp. (fungu, Chrysobalanaceae), *Vouacapoua americana* (bruinhart, Fabaceae), *Terminalia amazonia* (djindja udu, Combretaceae), *Eschweilera corrugata* (umabarklak, Lecythidaceae), *Eperua falcata* (walaba, Fabaceae), *Goupia glabra* (Goupiaceae), *Ceiba pentandra* (kankantri, Malvaceae), *Elizabetha princeps* (Fabaceae) and different species of Burseraceae. The composition of this seasonally flooded forest seemed to resemble to some extent the composition of the high tropical rainforest on dryland (terra firme).

   We sampled along the Kutari River downriver from our first camp site (towards the Aramatau River) and noted the forest at the river margin changed in terms of species composition. In addition to the aforementioned tree species found along the Wioemi Creek and the Kutari River, trees of *Virola* sp. (babun udu, Myristicaceae), *Triplaris surinamensis*, *Ceiba pentandra*, and palm (*Attalea maripa*, *A. microcarpa*) appeared in the forest. We found a similar species composition along the Sipaliwini River, despite the higher riverbanks.

Downstream from the confluence of the Kutari and Aramatau Rivers, we found stretches of floodplain forest with a swampy character. *Astrocaryum sciophilum* did not occur in this area, suggesting that soils could be wet throughout the year. The forest composition was dominated by trees of *Virola* sp., *Alexa wachenheimii*, and *Bixa orellana* (kusuwe, Bixaceae). Along the Sipaliwini River this floodplain forest only occurred where the riverbanks were low.

3. **(Seasonally flooded) palm swamp forest.** Close to the Kutari River camp, we observed patches of *Euterpe oleracea* (pina palm, Arecaceae) swamp forest, with occasional *Geonoma baculifera* (taspalm, Arecaceae). This swamp forest was slowly flooded during our stay by a nearby overflowing creek. At the hinterland of the Sipaliwini River camp, we found a stretch of swamp forest with a dense cover of *Geonoma baculifera*. This swamp forest was also close to a creek, and could be seasonally inundated.

4. **High tropical lowland rainforest on dryland (terra firme).** This was one of the most dominant forest types in the Kwamalasamutu region, occurring at all RAP survey sites. The understory of the high tropical rainforest on dryland at the Kutari River (Site 1) was dense and dominated by *Astrocaryum sciophilum*. Soils at the Kutari site appeared to contain a higher proportion of loam and clay than sand compared to the other RAP sites. Some other frequently encountered species were *Vouacapoua americana*, *Bocoa viridiflora* (ijzerhart, Fabaceae), *Bocoa marionii*, *Croton matourensis* (tabakabron, Euphorbiaceae), *Protium* and *Tetragastris* sp. (Burseraceae), *Licania* sp., *Eschweilera* sp., *Sagotia racemosa* (zwarte taja udu, Euphorbiaceae), *Bixa orellana*, and several Meliaceae and Lauraceae species.

At Werehpai (Site 3), we encountered many creeks along the main trail to the Werehpai caves. The forest along this trail was diverse, shifting from secondary forest to swampy, low and open vegetation, to high forest over short distances. The high tropical rainforest was dominated by *Astrocaryum sciophilum*, *Eperua falcata*, *Apeiba petoumo*, *Alexa imperatricis*, *Licania* sp., *Carapa guianensis*, *Eschweilera* and *Lecythis* sp., *Protium* and *Tetragastris* sp., *Inga* sp., *Guarea grandifolia* (Meliaceae), and *Couratari stellata* (ingi pipa, Lecythidaceae).

5. **High Tropical forest on laterite/granite hills.** This forest type was especially dominant in the higher areas along the Sipaliwini River (Site 2), but occurred at Werehpai as well (Site 3). *Astrocaryum sciophilum* was dominant where soils were deep and the understory of the forest relatively open. Also common here were *Alexa imperatricis*, *Vouacapoua americana*, *Inga* sp., *Protium* and

Tetragastris sp., *Licania* sp., *Eschweilera* and *Lecythis* sp., *Carapa guianensis*, *Bocoa alterna*, and *Osteophloeum platyspermum*. On small granite hills with relatively shallow soils, we observed *Sterculia pruriens* (okro udu, Malvaceae), *Zanthoxylum rhoifolium* (pritjari, Rutaceae), *Lacmellea aculeata* (zwarte pritjari, Apocynaceae), *Hevea guianensis* (Euphorbiaceae), *Jacaranda copaia* (gubaja, Bignoniaceae), *Eschweilera corrugata*, *Sloanea* sp. (rafunyannyan, Elaeocarpaceae), *Cupania scrobiculata* (gawetri, Sapindaceae), *Licania ovalifolia* (santi udu, Chrysobalanaceae), and *Geissospermum sericeum* (bergi bita, Apocynaceae).

6. **Savannah (moss) forest.** In the surroundings of the Sipaliwini River site, we encountered savannah forest with a low canopy dominated by many lianas (e.g. Bignoniaceae) in higher areas where boulders and hard parent rock were at the surface, causing shallow soils (e.g., in a portion of plot Si3). At Werehpai we encountered some patches of this forest type along the main trail to the petroglyphs. Near the inselberg at Site 2, we found a small, narrow stretch of savannah forest with some moss coverage and grasses, and a low canopy forest with trees of *Neea* sp. (Nyctaginaceae) and Myrtaceae species. This savannah moss forest occurred at the edge of the open rock face (see below).

7. **Open rock (inselberg) vegetation.** At the hinterland of the Sipaliwini River site, we found a small inselberg rising up above the forest canopy. The vegetation of this inselberg was similar to the vegetation found on the Voltzberg in central Suriname. On the rocky outcrop itself, we observed *Furcraea* sp. (Agavaceae), *Neea* sp. (Nyctaginaceae), *Cissus verticillata* and *C. erosa* (Vitaceae), *Cochlospermum orinocense* (Cochlospermaceae), *Clusia* sp. (Clusiaceae), *Ernestia* sp. (Melastomataceae), and different species of Orchidaceae, Gesneriaceae, Myrtaceae, Poaceae, and Bromeliaceae.

8. **Secondary vegetation.** The camp at Werehpai (Site 3) was established on an old abandoned farm. The forest around this camp was a secondary forest dominated by *Cecropia* sp. and *Guadua* sp., (bamboo, Poaceae) and domesticated plants such as *Musa* sp. (bacove, Musaceae) and big trees of *Spondias mombin* (mope, Anacardiaceae). Along the Sipaliwini River we also observed open areas completely covered by vines such as *Dioclea virgata* (Fabaceae).

9. **Bamboo forest.** At all three study sites, and especially along the Sipaliwini River, patches of bamboo (*Guadua* sp.) occurred in the forest along the river edge. During reconnaissance flights, we were able to distinguish several square patches of bamboo, suggesting that bamboo had colonized areas previously cleared by humans. Bamboo was less common along Wioemi Creek.

Plant collections: new records and noteworthy plant species

In total we made 401 plant collections (see Appendix) with the following geographical distribution: 214 plant collections at the Kutari River (Site 1), 99 plant collections at the Sipaliwini River (Site 2), and 88 plant collections at Werehpai (Site 3). Of these 401 plant collections, 185 specimens were fertile; most of these specimens were collected during the general plant surveys. The rest of the collections were sterile and all originated from the plot inventories. To date, more than 90% of all the plant collections have been identified at least to family level (62 families), and almost 70% to genus (132 genera) and species level. We estimate that we have collected almost 240 species in total. A substantial part of the sterile collections and a small part of the fertile collections still need further identification to species level; therefore, it is possible that the total number of species in our sample will increase in the near future.

In the general plant collecting and plot surveys, we encountered eight plant species new to Suriname. *Bocoa marionii* (Fabaceae), a tree with unifoliolate leaves and white flowers, is a species just recently described based on two collections from the upper Essequibo River in Guyana (Aymard and Ireland 2010). We collected two fertile collections that are the first collections for Suriname and the fourth collections for the species as a whole (Ben Torke pers. comm.). Another new tree species for Suriname is *Bocoa alterna* (Fabaceae). This species was previously only collected from the Guiana Shield region in central Guyana and in Amapá (Brasil), but has an Amazonian distribution reaching into Peru, Bolivia, and the central and western parts of Brasil. Also new for Suriname is the tree species *Trichilia surumuensis* (Meliaceae). This species is thought to be endemic to the Roraima area of Guyana and Brasil, where it has been collected. For this reason it was placed on the IUCN Red List (see below). *Cupania macrostylis* is a species newly described (and still unpublished) by Pedro Acevedo of the Smithsonian Institution in the USA. We collected the second record of this species for Suriname. *Buchenavia parvifolia* (Combretaceae) is a tree species that was previously known in the Guianas from southern Guyana and French Guiana. The tree species is new for Suriname, but it has an Amazonian distribution in Brasil, Bolivia, Peru, and Ecuador. Another tree species new for Suriname is *Machaerium floribundum* (Fabaceae). It has a wide geographic distribution in Amazonia and Mesoamerica. The sixth new tree species for Suriname is *Licania granvillei* (Chrysobalanaceae), which also has an Amazonian distribution. We found two new species of herbs for Suriname: *Justicia sprucei* (Acanthaceae), previously known only from French Guiana, and *Dichorisandra hexandra* (Commelinaceae) which is new for Suriname according to the checklist of the Guianas (Funk et al. 2007), although Missouri Botanical Gardens mentions one collection of this species for Suriname (www.discoverlife.org). The geographical distribution of this species extends south of Amazonia and north into Mesoamerica.

We encountered several rare or noteworthy species for Suriname. We collected the second specimen of the tree species *Duguetia cauliflora* (Annonaceae) for Suriname (Paul Maas pers. comm.). We also collected what appears to be the second collection for Suriname of the liana *Byttneria cordifolia* (Malvaceae). The first collection for Suriname dated from 1926 and was collected by Gerold Stahel along the Upper Suriname River. Our specimen is the eighth collection for the Guianas in the NCB Naturalis. A collection of the tree species *Mosannona discolor* is the fourth collection for Suriname and the tenth collection for this species in general (Lars Chatrou pers. comm.). Noteworthy observations included an uncommon Myristicaceae tree species (*Osteophloeum platyspermum*) with amber-colored latex in the bark. The tree is called lapa lapa by the Trio people of Kwamalasamutu, and is quite common along the Sipaliwini River in high tropical rainforest on laterite/granite hills. On the banks of the Sipaliwini River we also found *Herrania kanukuensis* (Malvaceae), a small tree with a spectacular flower with long purple petals on the stem. This species is not common, and ours is the first specimen with flowers in the Guianas collection of the NCB Naturalis and the National Herbarium of Suriname. On the small inselberg close to the Sipaliwini camp we collected a tree, *Cochlospermum orinocense,* that is restricted to rocky outcrops and has large showy yellow flowers.

Plant species with a special status

During our fieldwork we recognized several plant species that are protected under Surinamese law, or have a special designation on the IUCN Red List or CITES.

We encountered six tree species listed on the IUCN Red List:

- *Aniba rosaeodora* Endangered (EN) A1d+2d ver 2.3 (1998)
- *Cedrela odorata* Vulnerable (VU) A1cd+2cd ver 2.3 (1998)
- *Corythophora labriculata* Vulnerable (VU) D2 ver 2.3 (1998)
- *Minquartia guianensis* Lower Risk / Near Threatened (LR/nt) ver 2.3 (1998)
- *Trichilia surumuensis* Endangered (EN) B1+2c (1998)
- *Vouacapoua americana* Critically Endangered (CR) A1cd+2cd ver 2.3 (1998)

We encountered three tree species protected under Surinamese law:

- *Aniba rosaeodora* (rozenhout)—also on CITES Appendix II
- *Dipteryx odorata* (tonka)
- *Manilkara bidentata* (boletri)

The taspalm (*Geonoma baculifera*) does not have protected status. However, the leaves of this small palm tree are used for roof thatch. In the northern regions of Suriname, populations of the taspalm have seriously declined in recent years, potentially causing local extinctions. In the forests of the Kwamalasamutu region, this palm species still appears to be relatively abundant.

Plot Inventories

In all plots combined, we found a total of 1849 individual trees belonging to 54 families, 119 genera, and approximately 250 species. The twenty most common species were: *Astrocaryum sciophilum* (174 individuals), *Alexa imperatricis* (105), *Pausandra martinii* (82), *Vouacapoua americana* (74), Lecythidaceae sp. (62), *Bocoa viridiflora* (61), *Protium* sp. (60), *Eperua falcata* (53), *Licania albiflora* (42), *Balizia pedicellaris* (39), *Carapa guianensis* (33), *Licania majuscula* (29), *Tetragastris altissima* (28), *Geissospermum sericeum* (26), *Guarea* OSB 1340 (25), *Lecythis corrugata* (23), *Inga* OSB 1338 (22), *Chrysophyllum argenteum* (20), *Iryanthera hostmannii* (20), and *Rheedia benthamiana* (17).

In terms of tree alpha diversity, the Fisher's alpha values were highest in the Kutari River plots (Ku1&2; Table 1). The Fisher's alpha values were lowest in the Werehpai plots (We5&6), and intermediate in the Sipaliwini River plots (Si3&4). The Kutari River plots could be classified as 'high tropical lowland rainforest on dryland' (vegetation type 4, above); and the Sipaliwini plots as 'high tropical rainforest on laterite/granite hills' (vegetation type 5, above). The Werehpai plots were mixed between these two forest types, but most resembled the 'high tropical rainforest on laterite/granite hills' forest type.

We compared the 0.1-ha and 1-ha plots with other datasets from Suriname to determine the floristic differences (beta diversity) between the forests of the Kwamalasamutu region and forests in northern Suriname. The three 0.1-ha plots from the Kwamalasamutu area were separated floristically from plots in northern Suriname (Fig. 1). At the same time, the three 0.1-ha plots from the Kutari River, Sipaliwini River and Werehpai were to some extent also floristically separated from each other. The same patterns were found in the 1-ha plot dataset comparison (Fig. 2). This suggests beta diversity is substantial between the plots from the Kwamalasamutu area and those from northern Suriname. It also suggests that beta diversity among the plots of the three RAP sites is substantial.

We used indicator species analyses to investigate the number and identity of the species that were responsible for the separation of plots into different floristic groups. Only 18% of the species in the 0.1-ha plot dataset had a significant indicator value, and the number of indicator species for the plots in the Kwamalasamutu area was low (ca. 5% of all species). Several of the indicator species occurred in low numbers in the plots in northern Suriname, or were known to have a wide distribution. However, the plots in the Kwamalasamutu area contained specific indicator species. Again the 1-ha plots showed similar results. In both the 0.1-ha and 1-ha plot comparisons, the number of individuals, the number of species, and the values of Fisher's alpha did not differ significantly between the Kwamalasamutu area and northern Suriname.

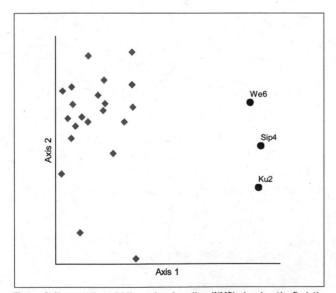

Figure 1. Non-metric multidimensional scaling (NMS) showing the floristic differences of the 0.1-ha plots of the Kutari River (Ku2), Sipaliwini River (Si4), and Werehpai (We6) with 0.1-ha forest plots in northern Suriname (diamond symbols). The first axis represents the most variation in floristic differences (52%) separating the 0.1-ha plots of the Kwamalasamutu surroundings from all other plots. The second axis represents 25% of the floristic variation mostly showing differences among the plots in northern Suriname, but also some floristic differences between the 0.1-ha plots of the three RAP sites.

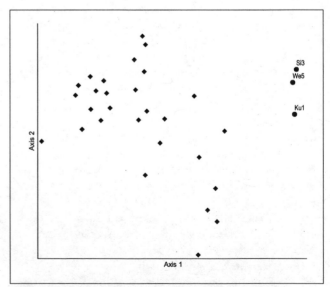

Figure 2. Non-metric multidimensional scaling (NMS) showing the floristic differences of the 1-ha plots of the Kutari River (Ku1), Sipaliwini River (Si3), and Werehpai (We5) with 1- ha plots of the forests in northern Suriname (diamond symbols). The first axis represents the most variation in floristic differences (55%) separating the 1-ha plots of the Kwamalasamutu surroundings from all other plots. The second axis represents 21% of the floristic variation mostly showing differences among the plots in northern Suriname, but also some floristic differences between the 1-ha plots of the three RAP sites.

DISCUSSION

The forests in the Kwamalasamutu region contradict the view that forests in southern Suriname are uniform. Instead, it is clear that the landscape is quite heterogeneous, forming a mosaic of different forest types over short geographic distances. Each of the three sites that we surveyed had a distinct species composition. At the same time, the alpha diversity of trees in our study plots differed substantially among the three sites. The Fisher's alpha values for the Kutari plots are comparable to the highest value calculated in Suriname to date (from a plot in the Lely Mountains; Bánki 2010). The values of Fisher's alpha for the Werehpai plots are in the range that is typically calculated for savannah forests, which are rather low values for tropical forests. And the Fisher's alpha values for the Sipaliwini plots are close to those found in plots on brown sands in the northern regions of Suriname (Bánki 2010). Although data from six plots are insufficient to extrapolate to the region as a whole, they do support the view that there are dramatic differences in tree alpha diversity over short geographic distances in the forests of the Kwamalasamutu region.

The differences in forest composition and diversity among plots from the three sites in the Kwamalasamutu area could be partly a result of the soils found in the plots. The forests in the Kutari plots stand on deep sandy to loamy soils, enabling a high canopy. In this forest type we also found the highest number of timber species. Nevertheless, in general, forests in the Kwamalasamutu area seem to have a low abundance and occurrence of commercial timber species. In the Sipaliwini and Werehpai plots, the species composition seemed to change instantly when the soils became shallow due to the hard parent rock underneath. These shallow soils could be relatively nutrient-poor and may have a reduced water retention capacity. Canopy height was reduced at those places where bedrock was close to the surface.

Compared to plots in northern Suriname, the plots in the Kwamalasamutu area had a distinct species composition, to some extent. We found a substantial number of tree species and two herbaceous species that were not previously recorded for Suriname. The forests in the Kwamalasamutu region are likely to resemble forests in southern Guyana; several of the new species for Suriname were previously known from there, including one species that was thought to be endemic to the Roraima area of Guyana and Brasil. Some of the other new species for Suriname had a much wider geographic distribution across Amazonia and some even beyond, to Mesoamerica. Despite the lower tree alpha diversity values from plots at the Sipaliwini River and especially at Werehpai, the number of new tree species records indicates these forests have a high natural value for Suriname.

The forests of the Kwamalasamutu region differed from other forests in Suriname in several ways. The seasonally flooded forest was quite extensive, with rivers and creeks actually flowing through the forests. Within the meandering rivers and creeks, the short swamp vegetation resembled coastal swamp vegetation types. Also of interest was the extent of semi-xerophytic forest types and open rock vegetation on small inselbergs that formed typical features in the landscape of the Kwamalasamutu region. Patchy bamboo forest was a conspicuous feature of the landscape. These forests were more extensive around Kwamalasamutu than elsewhere in Suriname.

CONSERVATION RECOMMENDATIONS

Based on our findings during this rapid assessment in the Kwamalasamutu region, we have formulated the following conservation recommendations:

The forests in the surroundings of Kwamalasamutu seem to have a high conservation value. This is expressed in the forest composition of the plots in the Kwamalasamutu surroundings that are to some extent different from forests in other plot datasets from northern Suriname. It is strengthened by the fact that we found eight species new for Suriname and a substantial number of rare species during the RAP survey. More field data on plant diversity and forest composition from the southern Suriname is very much needed to better determine the differences between these forests and those of northern Suriname, in terms of biomass and other ecosystem services.

The high conservation value is also demonstrated by the fact that the Kutari forest plots had one of the highest tree alpha diversity values ever recorded for Suriname. At the same time, the forests sampled at Werehpai had relatively low alpha diversity values relative to other forests in Suriname and the Guianas. Within the whole study area, the landscape was quite heterogeneous, with a high turnover of different vegetation types over short geographic distances. At sites where the landscape is a fine mosaic of different vegetation types, as is the case at Werehpai, conservation measures such as community managed protected areas are justified and should be promoted.

The majority of the forests in the surroundings of Kwamalasamutu are in a natural and healthy state. To ensure their continued health into the future, some matters need to be addressed in close cooperation with the Kwamalasamutu community. In the direct surroundings of Kwamalasamutu, the pressure of slash-and-burn agricultural methods is visible, especially on those forest types with a high tree alpha diversity. In the long run, this form of agriculture may not be sufficient to feed the population of Kwamalasamutu. At the same time, bamboo forest could spread over the area following human-induced disturbance, inhibiting the growth of other plant species. We therefore recommend that agricultural methods that better incorporate standing forests be tested in a community-supported approach in the Kwamalasamutu region.

REFERENCES

Aymard, G.A. & Ireland, H.E. 2010. A new species of *Bocoa* (leguminosae-Swartzieae) from the Upper Essequibo region, Guyana. Blumea 55: 18–20.

Bánki, O.S. 2010. Does neutral theory explain community composition in the Guiana Shield forests? Ph.D. dissertation, Universiteit Utrecht, Netherlands.

Dufrene, M. & Legendre, P. 1997. Species assemblages and indicator species: the need for a flexible asymmetrical approach. *Ecological Monographs*, 67, 345–366.

Fisher, R.A., Corbet, A.S., & Williams, C.B. 1943. The relation between the number of species and the number of individuals in a random sample of an animal population. Journal of Animal Ecology, 12, 42–58.

Funk, V., Hollowell, T., Berry, P., Kellof, C., & Alexander, S.N. 2007. Checklist of the Plants of the Guiana Shield (Venezuela: Amazonas, Bolivar, Delta Amacuro; Guyana, Surinam, French Guiana). *Contributions from the United States National Herbarium*, 55, 584.

Haripersaud, P.P. 2009. Collecting biodiversity. Ph.D. thesis, plant ecology and biodiversity group, institute of environmental biology, Utrecht University. 143pp.

Jansen-Jacobs, M.J. 1985. Flora of the Guianas. Royal Botanical Gardens Kew, Koelz Scientific Books.

Lindeman, J.C., and S.P. Moolenaar. 1959. Preliminary survey of the vegetation types of northern Suriname. Van Eeden Fonds, Amsterdam.

McCune, B. & Grace, J.B. 2002. Analysis of Ecological Communities. MjM Software Design, Gleneden Beach, Oregon, U.S.A.

McCune, B. & Mefford, M.J. 1999. PC-ORD. Multivariate Analysis of Ecological Data. MjM Software, Gleneden Beach, Oregon, U.S.A.

Pulle et al. 1932. Flora van Suriname. Rijksherbarium Utrecht, Van Eedenfonds & Bril Leiden, The Netherlands.

SPSS, Inc. 2007. SPSS 16.0 for windows; release 16.0.1 (Nov. 15, 2007).

Steyermark, J.A, Berry, P.E., Yatskievych, K., & Holst, B.K. 1995. Flora of the Venezuelan Guayana. Missouri Botanical Garden, St Louis, Missouri, USA.

ter Steege, H., Pitman, N.C.A., Phillips, O.L., Chave, J., Sabatier, D., Duque, A., Molino, J.-F., Prevost, M.-F., Spichiger, R., Castellanos, H., von Hildebrand, P., & Vasquez, R. 2006. Continental-scale patterns of canopy tree composition and function across Amazonia. *Nature*, 443, 444–447.

Appendix. List of plants collected on the Kwamalasamutu RAP survey. Numbers indicate number of specimens collected from each survey site.

Family	Genus	Species	Trio Name	Kutari	Sipaliwini	Werehpai
Acanthaceae	*Justicia*	*sprucei*			1	
Amaranthaceae					1	
Amaranthaceae	*Cyathula*	*prostrata*				1
Anacardiaceae			mapatalu			1
Anacardiaceae	*Loxopterygium*	*sagotii*			1	
Annonaceae	*Bocageopsis*	*multiflora*	lasaj	2		
Annonaceae	*Duguetia*	*cauliflora*	ariwera	1		
Annonaceae	*Duguetia*	*riparia*			1	
Annonaceae	*Duguetia*			3	2	
Annonaceae	*Fusaea*	*longiflora*			1	
Annonaceae	*Mosannona*	*discolor*	onote		1	1
Annonaceae	*Pseudoxandra*	*lucida*		1		
Annonaceae	*Unonopsis*	*glaucopetala*				1
Annonaceae	*Unonopsis*	*guatterioides*		1		
Annonaceae	*Xylopia*					1
Apocynaceae				1		
Apocynaceae	*Cynanchum*	*blandum*		1		
Apocynaceae	*Geissospermum*	*sericeum*	wataki	1		
Apocynaceae	*Mesechites*	*trifida*		2		
Apocynaceae	*Odontadenia*	*macrantha*		2		
Apocynaceae	*Pacouria*	*guianensis*		1		
Apocynaceae	*Tabernaemontana*	*heterophylla*				1
Apocynaceae	*Tabernaemontana*	*undulata*			1	
Araceae	*Heteropsis*	*flexuosa*				1
Asteraceae	*Mikania*	*cordifolia*		1		
Asteraceae	*Mikania*	*guaco*		1		
Bignoniaceae				1		
Bignoniaceae	*Arrabidaea*	*tuberculata*			1	
Bignoniaceae	*Cydista*	*aequinoctialis*		1		1
Bignoniaceae	*Martinella*	*obovata*		1		
Bignoniaceae	*Memora*	*schomburgkii*		1	1	
Bixaceae	*Bixa*	*orellana*	Wiseima, Kanawirike	2		
Boraginaceae	*Tournefortia*	*cuspidata*			1	
Burseraceae						1
Burseraceae	*Protium*			3		
Burseraceae	*Protium*	*aracouchini*	srisrituri	1		
Burseraceae	*Protium*	*heptaphyllum*			1	
Burseraceae	*Protium*	*robustum*			1	
Burseraceae	*Protium*	*trifoliolatum*				2
Burseraceae	*Tetragastris*	*altissima*	Arita	3	1	1
Burseraceae	*Tetragastris*	*hostmannii*	Arita	1		
Cecropiaceae	*Pourouma*	*bicolor*	alawata puruma	2		
Cecropiaceae	*Pourouma*	*minor*	wedinaiennu	1		

table continued on next page

Family	Genus	Species	Trio Name	Kutari	Sipaliwini	Werehpai
Celastraceae						1
Celastraceae	Cheiloclinium	cognatum		1		
Chrysobalanaceae	Couepia	caryophylloides		1		
Chrysobalanaceae	Hirtella	racemosa	merimeri	3	1	
Chrysobalanaceae	Licania			3		
Chrysobalanaceae	Licania	albiflora	paripo	1		
Clusiaceae	Rheedia	benthamiana	kaiquiennorepereke	1		
Clusiaceae	Tovomita			1	2	
Cochlospermaceae	Cochlospermum	orinocense			1	
Combretaceae						1
Combretaceae	Buchenavia	parvifolia	otaima	1		
Combretaceae	Combretum			1		
Combretaceae	Combretum	laxum		2	1	1
Combretaceae	Combretum	rotundifolium		1		
Combretaceae	Terminalia	dichotoma				1
Commelinaceae	Dichorisandra	hexandra	Lue	2		
Convolvulaceae	Evolvulus	alsinoides			1	
Convolvulaceae	Ipomoea	batatoides		1	1	
Convolvulaceae	Ipomoea	tiliacea			1	
Cyperaceae	Diplasia	karataefolia		1		
Cyperaceae	Rynchospora					1
Dichapetalaceae	Tapura	amazonica	awaima	1		
Elaeocarpaceae	Sloanea			2		
Elaeocarpaceae	Sloanea	grandiflora				1
Euphorbiaceae				1		1
Euphorbiaceae	Conceveiba	guianensis	kananamang ipaimu		1	1
Euphorbiaceae	Croton	matourensis	kuapehe	1		
Euphorbiaceae	Croton	sipaliwinensis		1		
Euphorbiaceae	Omphalea	diandra			1	
Euphorbiaceae	Pausandra	martinii	masiwewarito	1		1
Fabaceae				3	2	1
Fabaceae	Alexa	imperatricis	otoima, hotojarang	5	1	
Fabaceae	Andira	surinamensis		1		
Fabaceae	Bocoa	alterna				1
Fabaceae	Bocoa	marionii	pade	4		
Fabaceae	Bocoa	viridiflora	kutari	1		
Fabaceae	Cynometra	marginata			2	
Fabaceae	Dalbergia	riedelii		1	1	
Fabaceae	Dioclea	virgata		1		
Fabaceae	Eperua	rubiginosa		1		
Fabaceae	Hydrochorea	corymbosa		1		1
Fabaceae	Inga			1	1	2
Fabaceae	Inga	bourgonii		1		1
Fabaceae	Inga	disticha		1		

table continued on next page

Family	Genus	Species	Trio Name	Kutari	Sipaliwini	Werehpai
Fabaceae	*Inga*	*nobilis*		2		1
Fabaceae	*Inga*	*sertulifera*				1
Fabaceae	*Machaerium*	*floribundum*		1	1	
Fabaceae	*Machaerium*	*leiophyllum*		1		
Fabaceae	*Macrolobium*	*acaciifolium*			1	
Fabaceae	*Peltogyne*	*venosa*		1		
Fabaceae	*Senna*	*bicapsularis*			1	
Fabaceae	*Senna*	*quinquangulata*				1
Fabaceae	*Stylosanthes*	*hispida*			1	
Fabaceae	*Swartzia*			1		1
Fabaceae	*Swartzia*	*arborescens*		1		
Fabaceae	*Swartzia*	*benthamiana*		1		2
Fabaceae	*Swartzia*	*grandifolia*		1		
Fabaceae	*Swartzia*	*guianensis*			1	
Fabaceae	*Tachigali*	*paniculata*		1		
Fabaceae	*Vouacapoua*	*americana*	wacapu	1		
Fabaceae	*Zygia*	*inaequalis*		1		
Fabaceae	*Zygia*	*latifolia*			1	1
Fabaceae	*Zygia*	*racemosa*	krikriia	2		
Goupiaceae	*Goupia*	*glabra*			1	
Humiriaceae	*Humiria*	*balsamifera*	weikepauudu	1		
Lauraceae				8	1	1
Lauraceae	*Kubitzkia*	*mezii*			1	
Lauraceae	*Licaria*	*cannella*			1	
Lecythidaceae					1	1
Lecythidaceae	*Corythophora*	*labriculata*		1		
Lecythidaceae	*Eschweilera*			1		
Lecythidaceae	*Eschweilera*	*pedicellata*		1		
Lecythidaceae	*Eschweilera*	*sagotiana*				1
Lecythidaceae	*Eschweilera*	*subglandulosa*			1	
Lecythidaceae	*Gustavia*	*augusta*				1
Lecythidaceae	*Gustavia*	*hexapetala*	kanaimanagpataimu, patunailophue	2		
Lecythidaceae	*Lecythis*	*corrugata*	tuhaima	1		
Lecythidaceae	*Lecythis*	*poiteaui*	adiwera	1		
Loganiaceae	*Strychnos*	*guianensis*		1		
Malphigiaceae	*Hiraea*	*faginea*		1		1
Malpighiaceae	*Heteropterys*	*macrostachya*		2		
Malpighiaceae	*Stigmaphyllon*	*sinuatum*		2		
Malvaceae	*Apeiba*	*albiflora*		1		
Malvaceae	*Apeiba*	*petoumo*	mukete		1	
Malvaceae	*Byttneria*	*cordifolia*			1	
Malvaceae	*Herrania*	*kanukuensis*				1
Malvaceae	*Hibiscus*	*sororius*		1		

table continued on next page

Family	Genus	Species	Trio Name	Kutari	Sipaliwini	Werehpai
Malvaceae	*Melochia*	*ulmifolia*			1	
Malvaceae	*Quararibea*	*guianensis*		1		
Melastomataceae				1		
Melastomataceae	*Clidemia*			1		2
Melastomataceae	*Ernestia*				2	
Melastomataceae	*Miconia*					3
Melastomataceae	*Miconia*	*serrulata*		1		
Meliaceae					2	
Meliaceae	*Guarea*			8		
Meliaceae	*Guarea*	*guidonia*		1		
Meliaceae	*Trichilia*	*quadrijuga*			1	1
Meliaceae	*Trichilia*	*surumuensis*				1
Meliaceae	*Trichilia*				1	
Memecylaceae	*Mouriri*	*grandiflora*		1		
Menispermaceae				1		
Moraceae	*Bagassa*	*guianensis*			1	
Moraceae	*Brosimum*	*lactescens*			1	
Moraceae	*Brosimum*	*parinarioides*				1
Moraceae	*Brosimum*	*rubescens*				1
Moraceae	*Ficus*	*pertusa*		1		
Moraceae	*Ficus*	*trigona*		1		
Moraceae	*Helicostylis*	*tomentosa*	huhwe	1		
Moraceae	*Pseudolmedia*	*laevis*	mapanu	1		1
Moraceae	*Trymatococcus*	*amazonicus*				1
Myristicaceae	*Iryanthera*			1		1
Myristicaceae	*Iryanthera*	*hostmannii*	ponikrima	1		
Myristicaceae	*Osteophloeum*	*platyspermum*	lapalapa		1	
Myristicaceae	*Virola*			2	1	
Myrsinaceae	*Stylogyne*	*atra*				1
Myrtaceae				10	6	6
Myrtaceae	*Campomanesia*	*aromatica*				1
Myrtaceae	*Eugenia*		phumaime	1		
Myrtaceae	*Psidium*	*acutangulum*		3		1
Nyctaginaceae					1	
Ochnaceae				1		
Onagraceae	*Ludwigia*			2		
Opiliaceae	*Agonandra*	*silvatica*	alukaw	1	1	
Piperaceae	*Piper*			2		
Poaceae				1	2	2
Polygalaceae	*Securidaca*			1		1
Pteridophyte						2
Putranjivaceae	*Drypetes*	*variabilis*	tokirimang		1	1
Quiinaceae	*Lacunaria*	*crenata*		1		
Rhabdodendraceae	*Rhabdodendron*	*amazonicum*	payfayoinapiru	1		

table continued on next page

Family	Genus	Species	Trio Name	Kutari	Sipaliwini	Werehpai
Rubiaceae					3	5
Rubiaceae	Genipa	americana			1	
Rubiaceae	Palicourea	guianensis	alugeluge, pyia pyiama		1	
Rubiaceae	Posoqueria	longiflora		1		
Salicaceae				1		
Salicaceae	Casearia			1		
Sapindaceae	Cupania	macrostylis		1		
Sapindaceae	Cupania	scrobiculata	mapanu			1
Sapindaceae	Matayba			2		
Sapindaceae	Matayba	arborescens		1		
Sapindaceae	Matayba	camptoneura		1		
Sapindaceae	Paullinia	capreolata			1	
Sapindaceae	Paullinia	trilatera		1		
Sapindaceae	Talisia	sylvatica		1		
Sapindaceae	Toulicia	pulvinata			1	
Sapindaceae	Toulicia	elliptica		1		
Sapindaceae	Vouarana	guianensis			1	
Sapotaceae				7	10	2
Sapotaceae	Chrysophyllum	argenteum	tumuri	1		
Sapotaceae	Pouteria		awaribalata	1		
Sapotaceae	Pouteria	guianensis	kununima	1		
Siparunaceae	Siparuna	cuspidata	idakaipu	4		
Siparunaceae	Siparuna	decipiens	kandadeennu		1	
Solanaceae	Brunfelsia	guianensis				1
Solanaceae	Cestrum	latifolium			1	
Solanaceae	Schwenkia	grandiflora		1		
Solanaceae	Solanum	pensile		1		
Ulmaceae	Ampelocera	edentula		1		
Violaceae	Corynostylis	arborea				1
Violaceae	Rinorea			2		
Viscaceae	Phoradendron			1		
Vitaceae	Cissus	erosa	Napokaima	1	1	
Vitaceae	Cissus	verticillata		1	1	
Undetermined				13	9	11

Chapter 3

Odonata (dragonflies and damselflies) of the Kwamalasamutu region, Suriname

Natalia von Ellenrieder

SUMMARY

Odonata were studied during a Rapid Assessment Program (RAP) survey of the Kwamalasa-mutu area in SW Suriname. Ninety-four species, representing one-third of the species known from Suriname, were registered at forest rivers, streams, and swamps; in particular 57 species were found at the Kutari River Site (Camp 1), 52 at the Sipaliwini River Site (Camp 2), and 65 at the Werehpai Site (Camp 3). Fourteen species represent new records for Suriname, of which four, belonging to the genus *Argia*, are new to science, and five represent first records of a species at a new locality since their original descriptions, increasing considerably their known extent of occurrence. The results indicate a healthy watershed and well preserved forest at all three sites; if forest cover and stream morphology are maintained in the area, the present odonate assemblages are expected to persist.

INTRODUCTION

Dragonflies and damselflies (Order Odonata) are widespread and abundant in all continents with the exception of Antarctica, with centers of species richness occurring in tropical forests. As larvae they live in aquatic habitats and use a wide range of terrestrial habitats as adults. Larvae are sensitive to water quality and habitat morphology such as bottom substrate and aquatic vegetation structure, and adult habitat selection is strongly dependent on aerial vegetation structure, including degrees of shading. As a consequence dragonflies show strong responses to habitat change such as thinning of forest and increased erosion. Common species prevail in disturbed or temporary waters, whereas pristine streams, seepage, and swamp forests house an array of more vulnerable, often localized species. Thus odonates are useful for monitoring the overall biodiversity of aquatic habitats and have been identified as good indicators of environmental health (Corbet 1999; Kalkman et al. 2008). Due to their low species numbers relative to other insects (about 5,700 species worldwide) they also constitute an ideal target group for a Rapid Assessment Program because it is feasible to fully document their species diversity for a particular area in a relatively short period of time. This is the first instance where odonates were included in a RAP survey in South America. The taxonomy of the odonates from Suriname is relatively well known in general compared to that of other South American countries, since two odonate specialists devoted over 60 years of continuous research to its study (Geijskes 1931, 1943, 1946, 1954, 1959, 1976, 1986; Belle 1963, 1966a, 1966b, 1970, 1984, 1992, 2002). However, no published data regarding regional distribution or particular ecological requirements of the odonates from Suriname exists at this moment, and the Kwa-malasamutu area has never before been sampled for odonates.

METHODS AND STUDY SITES

Odonata species from the Kwamalasamutu area, in the Sipaliwini District of SW Suriname were studied by applying search-collecting methods. Odonates were surveyed from 19–24 August 2010 in the area surrounding the Kutari River (Camp 1: N 02°10'27", W 056°54'25", 263 m); from 28–31 August in the area adjacent to the Sipaliwini River (Camp 2: N 02°19'48", W 056°39'20", 264 m); and from 2–7 September in the surroundings of Werehpai (Camp 3: N 02°21'45", W 056°41'54", 252 m). Odonates were also recorded at Iwana Samu (N 02°21'46", W 056°45'18", 255 m) on 28 August, and at a vegetated ditch in Kwamalasamutu (N 02°21'17", W 056°47'11"W, 211 m) on 8 September. Searching, photographing, and collecting of adult odonates with an entomological aerial net was carried out around each camp, in terra firme forest along trails and in clearings, in forest swamps, along forest streams, and along rivers from a boat. All aquatic forest habitats were at least partially shady and usually devoid of aquatic plants, with an enclosed canopy cover, and only in the larger creeks and rivers did sun penetrate the forest canopy creating sunspots and larger continuous sunny areas at or near ground level during midday. The ditch at Kwamalasamutu was the only aquatic environment found which was fully exposed to the sun and provided with abundant aquatic and riparian vegetation. Presence/absence information of species was recorded and relative abundance for each species was noted as rare (1–3 individuals seen), frequent (4–20 individuals seen), or common (21–50 individuals seen). Collected specimens are deposited at the California State Collection of Arthropods and at the National Zoological Collection of Suriname.

Total species richness expected for the area was calculated using first-order jackknife and Chao 2 estimators. Composition of odonate communities from the three camps was compared using percentage complementarity (a measurement of distinctness or dissimilarity; Colwell & Coddington 1994).

RESULTS

Overall, 45 odonate genera belonging to 10 families were collected at the three sites, with a total of 94 species. These represent one-third of the total number of odonate species reported for Suriname (282 species according to Belle 2002). In particular, 10 families, 31 genera, and 57 species were collected at the Kutari site; 10 families, 28 genera, and 52 species at the Sipaliwini site; and 10 families, 34 genera, and 65 species at the Werehpai site. The odonates recorded at Kwamalasamutu and Iwana Samu represented 18 species in 13 genera and four families. Appendix A lists all species detected during the RAP survey and their relative abundances at each survey site.

The first-order jackknife estimator for the total number of odonates to be expected in this area was 120.3 species, and the Chao2 estimator was 137.23 species.

Werehpai was the richest site in odonate genera and species, and hosted 18 species not found at the other two camps: *Acanthagrion chacoense*, *Perilestes attenuatus*, *Brechmorrhoga praedatrix*, *Orthemis coracina* (along rivers), *Archaeogomphus nanus*, *Progomphus brachycnemis*, *Macrothemis ludia* (in forest creeks and streams), *Metaleptobasis quadricornis*, *M. mauritia* (at forest swamps), *Gynacantha gracilis*, *G. klagesi*, *G.* sp., *Misagria calverti*, *M. parana*, *Orthemis anthracina*, *Orthemis cultriformis*, *Uracis fastigiata*, and *U. siemensi* (along forest trails and in forest clearings). Twelve species were found only at the Kutari site: *Acanthagrion indefensum*, *Argia fumigata*, *Ebegomphus demerarae*, *Macrothemis delia* (at rivers), *Mnesarete cupraea*, *Macrothemis* sp. (in creeks and streams), *Argia* sp. 3, *Psaironeura tenuissima*, *Argyrothemis argentea* (in forest swamps), and *Mecistogaster ornata*, *Erythrodiplax castanea*, and *Gynothemis pumila* (along forest trails and clearings). Seven species were present only at the Sipaliwini site: *Phyllocycla ophis* (at rivers), *Neoneura mariana*, *Protoneura calverti*, *Elga leptostyla*, *Macrothemis hemichlora* (at forest creeks and streams), *Triacanthagyna ditzleri*, and *Macrothemis declivata* (along forest trails and clearings). Five of the species found at Kwamalasamutu (at a vegetated ditch) were unique to this site: *Miathyria simplex*, *Micrathyria artemis*, *Nephepeltia flavifrons*, *Oligoclada rhea*, and *Tauriphila argo*.

In terms of odonate community composition, the first and third camps were more dissimilar (complementarity of 51.8 %) than the second and third camps (complementarity of 48.6 %) or second and first camps (complementarity of 46.4 %; Table 1). Shared species usually showed different abundances at each one of the camps (i.e., many species common at one site were rare at another site; see Relative Abundance in Appendix A).

Four species of the genus *Argia* are new to science (*Argia* sp. 1, *A.* sp. 2, *A.* sp. 3, *A.* sp. 4). *Argia* is the most species-rich odonate genus in the New World, with 112 described species (Garrison et al. 2010). This genus shows its prevalence in all three sites being the richest in species (eight species total; four to eight species per camp). All of these new species are known also from collections outside of Suriname (Garrison pers. comm., Table 2), and are being described by Dr. Rosser W. Garrison (California Department of Food and

Table 1. Richness and percentage complementarity (number of species in common in parentheses) of odonate assemblages among Kutari, Sipaliwini, and Werehpai sites, Kwamalasamutu area, SW Suriname.

	Kutari	Sipaliwini	Werehpai
Species richness	57	52	65
Camp 2	46.4 (38)		
Camp 3	51.8 (40)	48.6 (40)	

Agriculture) as part of his ongoing taxonomic revision of the genus.

Another nine species were recorded from Suriname for the first time, and the discovery of five of these species (Figs. 2–5) represents also the first published record in the literature of a new locality since their original descriptions, increasing considerably their known extent of occurrence:

Perilestes gracillimus (page 17): Recorded from creeks in lowland Amazon forest from Peru (Kennedy 1941) and Brazil (Lencioni 2005).

Epipleoneura pereirai (Fig. 2): Previously known from rivers in lowland Amazon forest in Amapá and Pará States in Brazil (Machado 1964).

Neoneura angelensis (Fig. 3): Recently described (Juillerat 2007) from French Guiana.

Neoneura denticulata: Widely distributed in the lowland Amazon forest, from Venezuela and Brazil to Peru and Ecuador.

Elasmothemis rufa (Fig. 4): Recently described (De Marmels 2008) from Amazonas State in Venezuela.

Gynothemis pumila: Widely distributed across South America, from Colombia and Trinidad to Brazil and Peru.

Macrothemis ludia (Fig. 5): Known so far from Bolívar State in Venezuela (Belle 1987).

Micrathyria paruensis: Known from lowland Amazon forest in Venezuela and French Guiana.

Orthemis anthracina: lowland Amazonian rainforest in Venezuela (De Marmels 1989).

Orthemis coracina (Fig. 6): Known from lowland Amazonian forest in Ecuador (von Ellenrieder 2009).

None of the species found are endemic either to the study area or to Suriname. No odonates are listed on the CITES appendices. The conservation status of about one-quarter of the Neotropical species was recently assessed by the IUCN Odonate Specialist Group (Claustnitzer et al. 2009) including approximately one-fifth of the species found in the present study (Table 2). From these, most were assessed as Least Concern and two species, *Epipleoneura pereirai* and *Perilestes gracillimus*, as Data Deficient. Notes on the biology of the recorded species are provided in Appendix B.

Table 2. Odonates found in SW Suriname, Kwamalasamutu region: Habitat where found, data on known larvae, distribution, and conservation status according to IUCN criteria. Distribution=US: United States of America, ME: Mexico, GU: Guatemala, BE: Belize, ES: El Salvador, HO: Honduras, NI: Nicaragua, CR: Costa Rica, PA: Panama, CO: Colombia, VE: Venezuela, TR: Trinidad/Tobago, GY: Guyana, SU: Surinam, FR: French Guyana, BR: Brazil, EC: Ecuador, PE: Peru, BO: Bolivia, PY: Paraguay, UR: Uruguay, AR: Argentina. IUCN category= LC: Least Concern, DD: Data Deficient.

Species	Habitat	Larva described	Distribution	IUCN
Hetaerina caja dominula	river/creek	Geijskes 1943	ME, NI, CR, PA, CO, VE, TR, FR, EC, PE	-
Hetaerina moribunda	creek/trail	Geijskes 1943 by supposition	VE, GY, SU, FR, BR	-
Hetaerina mortua	creek	-	VE, GY, SU, FR, BR, PE	-
Mnesarete cupraea	creek	-	VE, GY, SU, FR, PE, BO	-
Acanthagrion ascendens	creek/ditch	Geijskes 1943	CO, VE, TR, GY, SU, FR, BR, EC, PE, BO	-
Acanthagrion chacoense	river	-	VE, SU, BR, PE, BO	LC
Acanthagrion indefensum	river	Geijskes 1943	VE, GY, SU, FR, BR	-
Acanthagrion rubrifrons	swamp	-	VE, GY, SU, FR, BR	-
Argia fumigata	river	-	VE, GY, SU, FR, BR	-
Argia insipida	river	Geijskes 1943	CR, CO, VE, TR, GY, SU, FR, BR	-
Argia oculata	trail/swamp	Limongi 1983 (1985)	ME, GU, BE, ES, HO, NI, CR, PA, CO, VE, TR, BR, EC, PE	-
Argia translata	river	Geijskes 1946, von Ellenrieder 2007	US, ME, GU, BE, ES, HO, NI, CR, PA, CO, VE, TR, SU, FR, EC, PE, AR	-
Argia sp. 1	trail/swamp	-	VE, GY, SU, FR, BR	-
Argia sp. 2	swamp	-	SU, FR	-
Argia sp. 3	swamp	-	SU, FR	-
Argia sp. 4	swamp	-	VE, SU, FR, BR	-
Inpabasis rosea	swamp	-	VE, SU, FR, BR	-
Metaleptobasis mauritia	swamp	-	TR, GY, SU, FR, BR	-
Metaleptobasis quadricornis	swamp	-	GY, SU, BR	-
Heliocharis amazona	creek	Geijskes 1986, Santos & Costa 1988	CO, EC, PE, BO, VE, GU, SU, FR, BR, PY, AR	-

table continued on next page

Table 2. *continued*

Species	Habitat	Larva described	Distribution	IUCN
Heteragrion ictericum	trail/creek/river	-	VE, GY, SU, FR, BR	-
Heteragrion silvarum	trail/creek	-	GY, SU, FR, BR	LC
Oxystigma williamsoni	trail/creek/river	-	VE, GY, SU, FR, BR	-
Perilestes attenuatus	river	Neiss & Hamada 2010	VE, SU, FR, BR, PE, BO	LC
Perilestes gracillimus	creek	-	SU, BR, PE	DD
Perilestes solutus	creek	-	VE, SU, BR	LC
Epipleoneura fuscaenea	river/trail	-	VE, GY, SU, FR	LC
Epipleoneura pereirai	river	-	SU, FR, BR	DD
Neoneura angelensis	river	-	SU, FR, BR	-
Neoneura denticulata	river/creek	-	VE, SU, BR, EC, PE	-
Neoneura joana	river	Geijskes 1954	VE, GY, SU, FR, BR	-
Neoneura mariana	creek	-	VE, GY, SU, FR	-
Neoneura myrthea	river	-	VE, GY, SU, FR, BO	-
Phasmoneura exigua	swamp	-	VE, GY, SU, FR, BR, PE	-
Protoneura calverti	creek	-	VE, TR, GY, SU, FR, BR	LC
Protoneura tenuis	creek	-	VE, TR, GY, SU, FR, BR, PE, BO	LC
Psaironeura tenuissima	swamp	-	GY, FR, BR, EC, PE	-
Mecistogaster lucretia	trail	-	CO, VE, GY, SU, FR, BR, EC, PE, AR	-
Mecistogaster ornata	trail	Ramirez 1995	ME, GU, ES, HO, NI, CR, PA, CO, VE, TR, SU, FR, BR, EC, PE, AR	LC
Microstigma anomalum	trail	-	SU, FR, BR, PE, BO	-
Gynacantha auricularis	trail	-	BE, NI, CR, VE, GY, SU, FR, BR, EC, PE, BO	-
Gynacantha gracilis	trail	Santos 1973a	GU, CR, PA, VE, GY, SU, FR, EC, PE, BO, AR	-
Gynacantha klagesi	trail	-	?	-
Gynacantha sp.	trail	-	VE, SU, FR, PE	-
Staurophlebia reticulata	river	Geijskes 1959	GU, BE, HO, NI, CR, PA, CO, VE, TR, GY, SU, FR, BR, EC, PE, PY, UR, AR	-
Triacanthagyna ditzleri	trail	-	ME, GU, BE, CR, PA, CO, VE, TR, GY, SU, FR, BR, EC, PE, BO	-
Aphylla sp.	river/creek	-	?	-
Archaeogomphu nanus	creek	Belle 1970	VE, SU, FR, BR	-
Ebegomphus demerarae	river	Belle 1966a, 1970	GY, SU	-
Phyllocycla ophis	river	Belle 1970	VE, GY, SU, FR, BR	-
Phyllogomphoides major	creek	Belle 1970 as *P. fuliginosus*	VE, GY, SU, FR, BR	-
Phyllogomphoides undulatus	river	Belle 1970 by supposition	VE, SU, FR, BR	-
Progomphus brachycnemis	river/creek	Belle 1966b	VE, SU, BR	LC
Argyrothemis argentea	swamp	Fleck 2003a	VE, GY, SU, FR, BR, PE	LC
Brechmorrhoga praedatrix	river	Fleck 2004	VE, TR, GY, SU, FR, BR, AR	LC
Diastatops pullata	river	Fleck 2003b	VE, GY, SU, FR, BR, EC, PE, BO, AR	LC
Dythemis multipunctata	creek	De Marmels 1982, Westall 1988	ME, GU, BE, ES, HO, NI, CR, PA, VE, TR, GY, SU, FR, BR, EC, PE, PY, AR	-
Elasmothemis cannacrioides	river	Westall 1988	ME, GU, BE, HO, CR, PA, CO, VE, TR, SU, FR, BR, EC, PE, AR	-
Elasmothemis rufa	river	-	VE, SU	-

table continued on next page

Table 2. *continued*

Species	Habitat	Larva described	Distribution	IUCN
Elga leptostyla	creek	De Marmels 1990, Fleck 2003b	PA, CO, VE, TR, GU, SU, FR, BR, EC, PE	-
Erythrodiplax castanea	swamp/clearing	-	GU, BE, CR, CO, VE, TR, GY, SU, FR, BR, EC, PE, BO, PY, AR	-
Erythrodiplax famula	clearing	-	CR, VE, TR, GY, SU, FR, BR, PE, AR	-
Erythrodiplax fusca	river/clearing/ditch	Santos 1967	ME, GU, BE, ES, HO, NI, CR, PA, CO, VE, TR, GY, SU, FR, BR, EC, PE, BO, PY, UR, AR	-
Fylgia amazonica lychnitina	swamp	De Marmels 1992	VE, GY, SU, FR, BR, EC, PE	LC
Gynothemis pumila	clearing	Fleck 2004	CO, VE, TR, GY, SU, FR, BR, PE	LC
Macrothemis declivata	clearing	-	SU, BR, PE, BO, PY, AR	-
Macrothemis delia	river	-	ME, GU, CR, VE, SU	-
Macrothemis hemichlora	creek	-	ME, GU, BE, ES, HO, CR, PA, CO, VE, TR, SU, FR, EC, PE, PY, AR	LC
Macrothemis ludia	clearing/creek	-	VE, SU	-
Macrothemis sp.	creek	-	?	-
Miathyria simplex	ditch	Limongi 1991	ME, GU, BE, HO, CR, PA, VE, TR, GY, SU, FR, BR, EC, PE	-
Micrathyria artemis	ditch	Santos 1972	VE, GY, SU, FR, BR, EC, PE, AR	LC
Micrathyria paruensis	creek	-	VE, SU, FR	-
Micrathyria spinifera	creek	-	CO, VE, TR, GY, SU, FR, BR, EC, PE, BO	LC
Misagria calverti	clearing	-	SU, FR, BR, PE	-
Misagria parana	clearing	-	VE, GY, SU, FR, BR, EC, PE	-
Nephepeltia flavifrons	ditch	-	CO, VE, SU, FR, BR, PE, PY, AR	-
Oligoclada abbreviata	river	Fleck 2003a	VE, GY, SU, FR, BR, EC, PE	LC
Oligoclada amphinome	swamp	-	VE, GY, SU, FR, BR	-
Oligoclada rhea	ditch	-	SU, FR, BR, BO	-
Oligoclada walkeri	creek	-	VE, TR, GY, SU, FR, BR, EC, PE	-
Orthemis aequilibris	clearing/river	Fleck 2003b	CR, PA, CO, VE, GY, SU, FR, BR, PE, BO, PY, AR	-
Orthemis anthracina	clearing	-	VE, SU	-
Orthemis coracina	river	-	EC, SU	-
Orthemis cultriformis	clearing/creek	-	CR, PA, CO, VE, TR, GY, SU, FR, BR, EC, PE, BO, PY, AR	-
Perithemis cornelia	creek	-	VE, SU, FR, BR, EC, PE, BO	LC
Perithemis lais	creek	Costa & Regis 2005	CO, VE, GY, SU, FR, BR, EC, PE, BO, PY, AR	LC
Perithemis mooma	creek/ditch	Santos 1973b, von Ellenrieder & Muzón 1999	ME, GU, BE, ES, HO, NI, CR, PA, CO, VE, TR, SU, FR, BR, EC, PE, PY, UR, AR	-
Perithemis thais	swamp/creek	Spindola, Souza & Costa 2001	CR, VE, TR, GY, SU, FR, BR, EC, PE, BO, AR	-
Tauriphila argo	ditch	Fleck, Brenk & Misof 2006	ME, GU, BE, HO, NI, CR, PA, VE, TR, GY, SU, FR, BR, EC, PE, BO, PY, AR	-
Uracis fastigiata	trail	-	ME, GU, HO, NI, CR, PA, CO, VE, TR, GY, SU, FR, BR, EC, PE, BO	-
Uracis infumata	trail	-	VE, GY, SU, FR, BR, PE, BO	-
Uracis ovipositrix	trail	-	VE, GY, SU, FR, BR, EC, PE	-
Uracis siemensi	trail	-	VE, SU, FR, BR, EC, PE	-

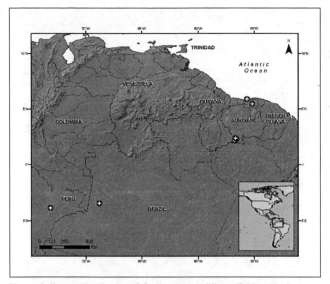

Figure 1. Known distribution of *Perilestes gracillimus* (Perilestidae).

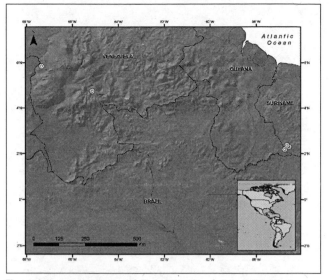

Figure 4. Known distribution of *Elasmothemis rufa* (Libellulidae).

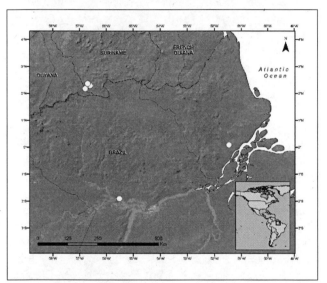

Figure 2. Known distribution of *Epipleoneura pereirai* (Protoneuridae).

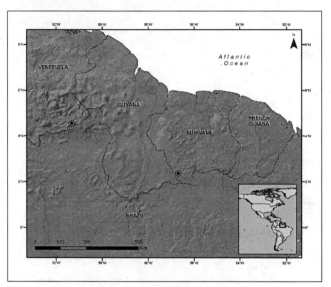

Figure 5. Known distribution of *Macrothemis ludia* (Libellulidae).

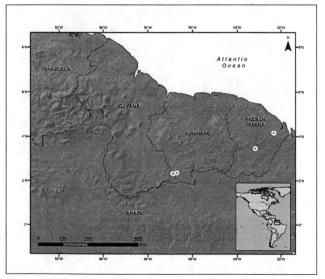

Figure 3. Known distribution of *Neoneura angelensis* (Protoneuridae).

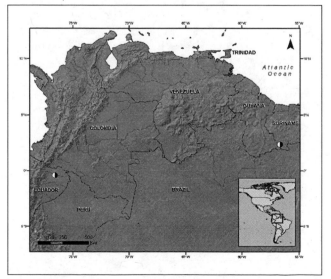

Figure 6. Known distribution of *Orthemis coracina* (Libellulidae).

DISCUSSION

The differences in odonate species composition among the three camps are probably due to slightly different qualities of the aquatic habitats sampled at each site, as a consequence of the different kinds of soils and vegetation which characterize the three camps (i.e., Kutari river and creeks were lower in oxygen content and pH; Kutari soils included loamy sands; soils were sandy and rocky at Werehpai).

Stagnant water bodies are comparatively scarce in the forest, and the presence of a vegetated ditch in Kwamalasamutu (a type of habitat not found at the camps) contributed five species restricted to lentic habitats. These species are all widespread, vagile species that inhabit open vegetated ponds throughout the lowland Amazon region (Table 2). A more thorough and extended study around Kwamalamasutu and Iwaana Saamu would most likely have recorded additional taxa characteristic of this type of environment in the Neotropics, and which are expected to occur in the area; i.e. species of *Lestes* (Lestidae), *Ischnura*, *Telebasis* (Coenagrionidae), *Erythemis*, *Tramea* (Libellulidae), and others. Diversity estimators corroborate this, indicating that another twenty to forty species are to be expected in this area.

Gynacantha sp. and *Macrothemis* sp. were represented by a single female each, and due to the poor taxonomic knowledge of the female sex in these two genera—unknown for many species—the specimens cannot presently be reliably assigned a specific name.

Aphylla sp. was represented by a larval exuviae. Although several species of this genus are known to occur in Suriname and adjacent lowland forests in the Guiana Shield, their larvae remain undescribed, and therefore it is not possible at this moment to identify the specimen to species.

Inpabasis rosea and *Metaleptobasis quadricornis* have been erroneously mentioned from Suriname in the literature as *I. eliasi* and *M. weibezahni*, respectively (Belle 2002). Belle (2002) also listed *Heteragrion melanurum* and *Microstigma maculatum* as present in Suriname (see Appendix A). According to De Marmels (1987), *H. melanurum* is a probable junior synonym of *Heteragrion silvarum*, an opinion which is shared in this study. Based on observed morphological variability of the diagnostic characters traditionally used to identify species of *Microstigma*, the three specific names currently used might represent only intraspecific variability. This was already suggested by Neiss et al. (2008) based on the apparent absence of diagnostic differences between the larvae of *M. maculatum* and *M. rotundatum*, and it is supported by the specimens studied here, which show an intergradation of characters attributed to *M. anomalum* and *M. maculatum*; thus the older of the two names (*M. anomalum*) is applied in this study.

CONSERVATION RECOMMENDATIONS

The diversity of odonate genera and species found in this study is typical of well preserved sites; most of the species found in the forest understory, creeks, and swamps in the three camps would not be present if the forest were disturbed.

Unlike birds, mammals, or butterflies, Odonata are largely unaffected by hunting or trade. However, many odonate species require closed canopy forest to maintain the appropriate vegetation structure they need as adults. Human activities such as deforestation and mining would most likely affect their occurrence in the area and produce a marked decrease in their diversity, since deforestation affects the vegetation structure needed by the adults, and subsequent alteration of water bodies by erosion and siltation would be detrimental for their larvae. Mining would lead to increased turbidity and siltation of streams, changing the substrate and reducing the habitat quality needed by odonate larvae. Claustnitzer et al. (2009) found that the threats are greater for forest species restricted to forest fragments, mountaintops, and island localities, whereas species inhabiting large forest blocks are usually at lower risk.

Therefore, the main conservation recommendation is to include an area as large as possible, encompassing at a minimum the three visited sites and intervening areas, as a legally protected nature preserve to prevent mining and logging activities and thus conserve the high diversity of odonate species found in this study. Species of particular conservation concern are the seldom-encountered lowland Amazon species including *Neoneura angelensis*, *Epipleoneura pereirai*, *Elasmothemis rufa*, *Macrothemis ludia*, and *Orthemis coracina*, all of which occur here (Figs. 2–6). If a nature preserve is not created and development activities do take place within the Kwamalasamutu region, it is recommended to establish broad buffer zones of undisturbed vegetation along rivers and creeks, in order to minimize the damage to the watershed and consequently to the odonate community.

Further biodiversity studies in the area are recommended to increase the knowledge of several poorly known and rare species that occur in these pristine forests (the larval stage of 61 % of species and habitat preferences of several of the species recorded are still unknown; see Table 2), and to gain knowledge about the possible seasonality (dry–rainy season species assemblages) of the odonate community of southwest Suriname. If further surveys are conducted, it is suggested to include some sites close to human settlements to provide the framework needed to assess which species are most affected by human disturbance and thus possibly identify indicator species of pristine environments for this area. Initial involvement of local communities in dragonfly and damselfly observation is encouraged, by providing them with educational color picture guides to the most common species in order to increase their appreciation and knowledge of this group, which could eventually be used for ecotourism and monitoring programs in the future.

REFERENCES

Bede, L.C., W. Piper, G. Peters, and A.B.M. Machado, 2000. Phenology and oviposition behaviour of *Gynacantha bifida* Rambur (Anisoptera: Aeshnidae). Odonatologica 29(4): 317–324.

Belle, J. 1963. Dragonflies of the Genus *Zonophora* with Special Reference to its Surinam Representatives. Stud. Fau. Surin. Guyan. 5(16): 60–69.

Belle, J., 1966a. Surinam dragon-files of the *Agriogomphus* complex of genera. Stud. Fau. Surin. Guyan. 8(29): 29–60.

Belle, J., 1966b. Surinam dragon-flies of the genus *Progomphus*. Stud. Fau. Surin. Guyan. 8(28): 1–28.

Belle, J., 1970. Studies on South American Gomphidae (Odonata) with special reference to the species from Surinam. Stud. Fau. Surin. Guyan. 11(43): 1–158

Belle, J., 1984. A synopsis of the South American species of *Phyllogomphoides*, with a key and descriptions of three new taxa (Odonata, Gomphidae). Tijdschr. v. Entomol. 127(4): 79–100.

Belle, J., 1987. Three new species of *Macrothemis* from northern South America (Odonata: Libellulidae). Zool. Med. Leiden 61(19): 287–294.

Belle, J., 1992. A revision of the South American species of *Aphylla* Selys, 1854 (Odonata: Gomphidae). Zool. Med. Leiden 66(12): 239–264.

Belle, J. 2002. Commented Checklist of the Odonata of Suriname. Odonatologica 31(1): 1–8.

Clausnitzer, V., V.J. Kalkman, M. Ram, B. Collen, J.E.M. Baillie, M. Bedjanic, W.R.T. Darwall, K.-D.B. Dijkstra, R. Dow, J. Hawking, H. Karube, E. Malikova, D. Paulson, K. Schütte, F. Suhling, R. Villanueva, N. von Ellenrieder, and K. Wilson. 2009. Odonata Enter the Biodiversity Crisis Debate: The First Global Assessment of an Insect Group. Biol. Conserv. 142: 1864–1869.

Colwell, R.K. and J.A. Codington. 1994. Estimating Terrestrial Biodiversity through Extrapolation. Phil. Trans. R. Soc. Lond. (B) 345: 101–118.

Corbet, P.S. 1999. Dragonflies - Behavior and Ecology of Odonata. Comstock Publishing Associates, Cornell University Press, Ithaca, NY.

Costa , J.M. and L.P.R.B. Regis. 2005. Description of the last instar larva of *Perithemis lais* (Perty) and comparison with other species of the genus (Anisoptera: Libellulidae). Odonatologica 34(1): 51–57.

De Marmels, J. 1982. Cuatro náyades nuevas de la familia Libellulidae (Odonata: Anisoptera). Bol. Entomol. Venez., N. S. 2(11): 94–101.

De Marmels, J. 1987. On the type specimens of some neotropical Megapodagrionidae, with a description of *Heteragrion pemon* spec. nov. and *Oxystigma caerulans* spec. nov. from Venezuela (Zygoptera). Odonatologica 16(3): 225–238.

De Marmels, J. 1989. Odonata or dragonflies from Cerro de la Neblina. Academia de las Ciencias Físicas, Matemáticas y Naturales, Caracas, Venezuela 25: 1–78.

De Marmels, J. 1990. Nine new Anisoptera larvae from Venezuela (Gomphidae, Aeshnidae, Corduliidae, Libellulidae). Odonatologica 19(1): 1–15.

De Marmels, J. 1992. Caballitos del Diablo (Odonata) de las Sierras de Tapirapecó y Unturán, en el extremo sur de Venezuela. Acta Biol. Venez. 14(1): 57–78.

De Marmels, J. 2007. Thirteen new Zygoptera larvae from Venezuela (Calopterygidae, Polythoridae, Pseudostigmatidae, Platystictidae, Protoneuridae, Coenagrionidae). Odonatologica 36(1): 27–51.

De Marmels, J. 2008. Three new libelluline dragonflies from southern Venezuela, with new records of other species (Odonata: Libellulidae). Int. J. Odonatol. 11(1): 1–13.

Fleck G. 2003a. Contribution à la connaissance des Odonates de Guyane française. Les larves des genres *Argyrothemis* Ris, 1911 et *Oligoclada* Karsch, 1889 (Insecta, Odonata, Anisoptera, Libellulidae). Ann. Naturhist. Mus. Wien, S. B 104: 341–352.

Fleck, G. 2003b. Contribution à la connaissance des Odonates de Guyane française: Notes sur des larves des genres *Orthemis, Diastatops* et *Elga* (Anisoptera: Libellulidae). Odonatologica 32(4): 335–344.

Fleck G. 2004. Contribution à la connaissance des Odonates de Guyane française: les larves des *Macrothemis pumilla* Karsch, 1889 et de *Brechmorhoga praedatrix* Calvert, 1909. Notes biologiques et conséquences taxonomiques (Anisoptera: Libellulidae). Ann. Soc. Entom. Fr. 40(2): 177–184.

Fleck G., M. Brenk, and B. Misof. 2006. DNA taxonomy and the identification of immature insect stages: the true larva of *Tauriphila argo* (Hagen 1869) (Odonata: Anisoptera: Libellulidae). Annales de la Société Entomologique de France (Nouvelle série) 42(1): 91–98.

Garrison, R.W., N. von Ellenrieder, and J.A. Louton. 2010. Damselfly genera of the New World. An Illustrated and Annotated Key to the Zygoptera. The Johns Hopkins University Press, Baltimore, xiv + 490 pp, + 24 color plates.

Geijskes, D.C. 1931. A New Species of *Oligoclada* (Odonata) from Trinidad B.W.Z. Ent. Berich. 8(178): 209–214.

Geijskes, D.C. 1943. Notes on Odonata of Surinam. IV. Nine new or little known zygopterous nymphs from the inland waters. Ann. Entomol. Soc. Amer. 36(2): 165–184.

Geijskes, D.C. 1946. Observations on the Odonata of Tobago, B.W.I.. Trans. Roy. Entomol. Soc. London 97(9): 213–235.

Geijskes, D.C. 1954. Notes on Odonata of Surinam VI. The nymph of *Neoneura joana* Will.. Entomol. News. 65(6): 141–144.

Geijskes, D.C. 1959. The aeschnine genus *Staurophlebia*. Stud. Fau. Surin. Guyan. 3(9): 147–172.

Geijskes, D.C. 1976. The genus *Oxystigma* Selys, 1862 (Zygoptera: Megapodagrionidae). Notes on Odonata of Surinam XIII. Odonatologica 5(3): 213–230.

Geijskes, D.C. 1986. The Larva of *Dicterias cothurnata* (Förster, 1906) (Zygoptera: Dicteriastidae). Odonatologica 15(1): 77–80.

González-Soriano, E. 1987. *Dythemis cannacrioides* Calvert, a libellulid with unusual ovipositing behaviour (Anisoptera). Odonatologica 16(2): 175–182.

IUCN 2001. IUCN Red List Categories: Version 3.1. Prepared by the IUCN Species Survival Commission. IUCN, Gland, Switzerland and Cambridge, UK.

Juillerat, L. 2007. *Neoneura angelensis* sp. nov. from French Guyana (Odonata: Protoneuridae). Int. J. Odonatol. 10(2): 203–208.

Kalkman, V.J., V. Clausnitzer, K.-D. Dijkstra, A. G. Orr, D.R. Paulson, and J. van Tol. 2008. Global Diversity of Dragonflies (Odonata) in Freshwater. Hydrobiologia 595: 351–363.

Kennedy, C.H. 1941. Perilestinae in Ecuador and Peru: revisional notes and descriptions (Lestidae: Odonata). Ann. Entomol. Soc. Amer. 34(3): 658–688.

Lencioni, F.A.A. 2005. Damselflies of Brazil. An illustrated identification guide. 1 - The non-Coenagrionidae families. All Print Editora iv + 324 pp.

Limongi, J. 1983 (1985). Estudio morfo-taxonómico de nayades en algunas especies de Odonata (Insecta) en Venezuela (I). Mem. Soc. Cs. Ns. "La Salle" 43(119): 95–117.

Limongi, J. 1991. Estudio morfo-taxonómico de náyades de algunas especies de Odonata (Insecta) en Venezuela (II). Memorias de la Sociedad de ciencias naturales "La Salle" 50(133–134): 405–420.

Louton, J.A., R.W. Garrison and O.S. Flint, 1996. The Odonata of Parque Nacional Manu, Madre de Dios, Peru: natural history, species richness and comparisons with other Peruvian sites, pp. 431–449. In: Wilson, D.E. and A. Sandoval (eds.), Manu, the biodiversity of southeastern Peru. Smithsonian Institution.

Machado, A.B.M. 1964. Duas novas Epipleoneuras dos Rios Paru de oeste e Amapari (Odonata- Protoneuridae). Bol. Mus. Paraense Emilio Goeldi, Zool. 51: 1–13.

Neiss, U.G., F.A.A. Lencioni, N. Hamada, and R.L. Ferreira-Keppler. 2008. Larval redescription of *Microstigma maculatum* Hagen in Selys, 1860 (Odonata: Pseudostigmatidae) from Manaus, Amazonas, Brazil. Zootaxa 1696: 57–62.

Neiss, U.G. and N. Hamada. 2010. The larva of *Perilestes attenuatus* Selys, 1886 (Odonata: Perilestidae) from Amazonas, Brazil. Zootaxa 2614: 53–58.

Ramírez, A., 1995. Descripción e historia natural de las larvas de odonatos de Costa Rica. IV. *Mecistogaster ornata* (Rambur, 1842) (Zygoptera, Pseudostigmatidae). Bull. Amer. Odonatol. 3(2): 43–47.

Santos, N.D. 1967. Notas sôbre a ninfa de *Erythrodiplax connata fusca* (Rambur, 1842) Brauer, 1868 (Odonata, Libellulidae). Atas Soc. Biol. Rio de Janeiro 10(6): 145–147.

Santos, N.D. 1969. Contribuição ao conhecimento da fauna do estado da Guanabara. 70--Descrição da ninfa de *Perilestes fragilis* Hagen in Selys, 1862 e notas sobre o imago (Odonata: Perilestidae). Atas Soc. Biol. Rio de Janeiro 12(5–6): 303–304.

Santos, N.D. 1972. Contribuição ao conhecimento fauna do estado da Guanabara e arredores. 80 - Descrição da ninfa de *Micrathyria artemis* (Selys ms.) Ris, 1911 (Odonata: Libellulidae). Atas Soc. Biol. Rio de Janeiro 15(3): 141–143.

Santos, N.D. 1973a. Contribuição ao conhecimento da fauna da Guanabara e arredores 82 - Descrição da ninfa de *Gynacantha gracilis* (Burmeister, 1839) Kolbe, 1888 (Aeshnidae: Odonata). Atas Soc. Biol. Rio de Janeiro 16(2–3): 55–57.

Santos, N.D. 1973b. Contribuição ao conhecimento da fauna do estado da Guanabara e arredores. 84 - Descrição da ninfa de *Perithemis mooma* Kirby, 1889 (Odonata: Libellulidae). Atas Soc. Biol. Rio de Janeiro 16(2–3): 71–72.

Santos, N.D. and J.M. Costa. 1988. The larva of *Heliocharis amazona* Selys, 1853 (Zygoptera: Heliocharitidae). Odonatologica 17(2): 135–139.

Spindola, L.A., L.O.I. Souza, and J.M. Costa. 2001. Descrição da larva de *Perithemis thais* Kirby, 1889, com chave para identificação das larvas conhecidas do gênero citadas para o Brasil (Odonata, Libellulidae). Bol. Mus. Nac., Rio de Janeiro, N.S. 442: 1–8.

von Ellenrieder, N. 2007. The larva of *Argia joergenseni* Ris (Zygoptera: Coenagrionidae). Odonatologica 36(1): 89–94.

von Ellenrieder, N. 2009. Five new species of *Orthemis* from South America (Odonata: Libellulidae). Int. J. Odonatol. 12(2): 347–381, pl. VII.

von Ellenrieder, N. & J. Muzón. 1999. The Argentinean species of the genus *Perithemis* Hagen (Anisoptera: Libellulidae). Odonatologica 28(4): 385–398.

Wasscher, M.T. 1990. Reproduction behaviour during heavy rainfall of *Oxystigma williamsoni* Geijskes in Surinam (Zygoptera: Megapodagrionidae). Notul. odonatol. 3(5): 79–80.

Westall, Jr., M.J. 1988. *Elasmothemis* gen. nov., a new genus related to *Dythemis* (Anisoptera: Libellulidae). Odonatologica 17(4): 419–428.

Westfall, Jr., M.J. and M.L. May. 2006. Damselflies of North America. Revised edition. Scientific Publishers, Inc. xii + 502 pp.

Williamson, E.B. 1918. A collecting trip to Colombia, South America. Univ. Mich. Mus. Zool., Misc. Pub. 3: 1–24.

Plate I. a: Male of *Hetaerina caja dominula* (Odonata, Calopterygidae) along a forest trail in Sipaliwini Camp.
b: Male of *Hetaerina moribunda* (Odonata, Calopterygidae) at a forest stream in Werehpai Camp. Both photographed by N. von Ellenrieder.

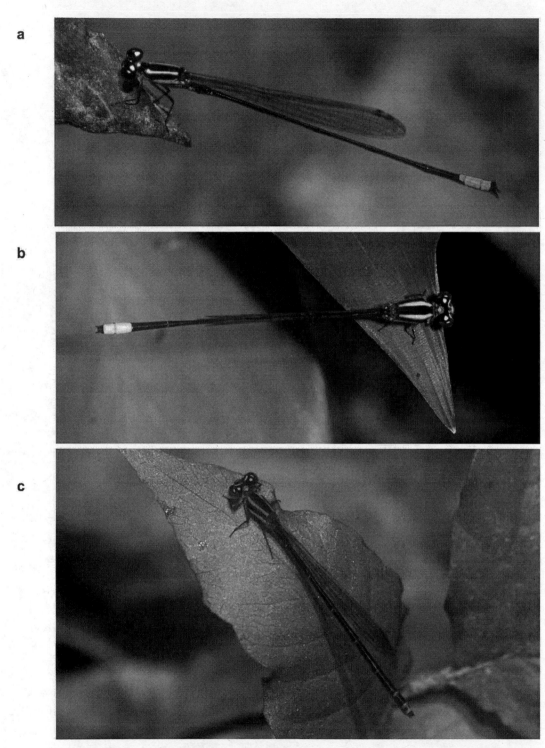

Plate II. a: Male of *Acanthagrion ascendens* (Odonata, Coenagrionidae) at a forest stream. **b**: Male, and **c**: Female, of *Acanthagrion rubrifrons* (Odonata, Coenagrionidae) at a forest swamp. All photographed at Sipaliwini Camp by N. von Ellenrieder.

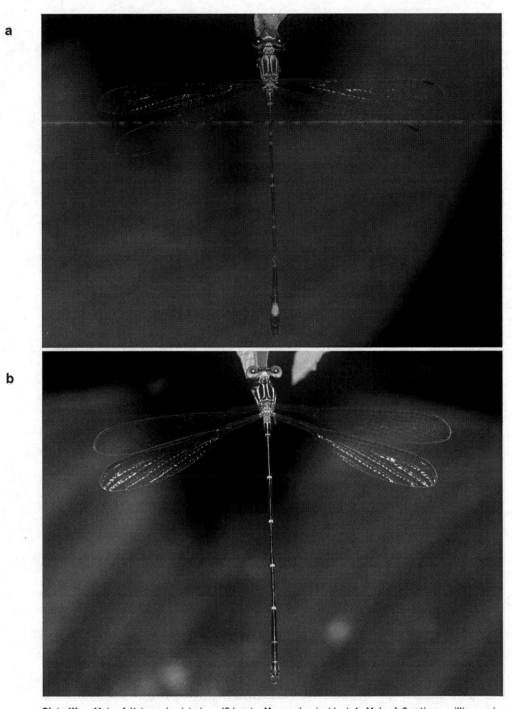

Plate III. a: Male of *Heteragrion ictericum* (Odonata, Megapodagrionidae). **b**: Male of *Oxystigma williamsoni* (Odonata, Megapodagrionidae). Both photographed at a forest creek in Werehpai Camp by N. von Ellenrieder.

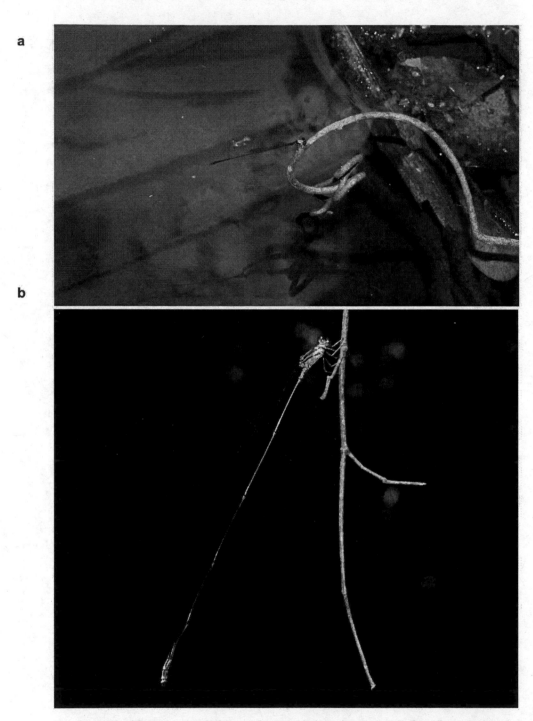

Plate IV. a: Male of *Protoneura tenuis* (Odonata, Protoneuridae) at a forest stream in Kutari Camp. **b**: Male of *Mecistogaster lucretia* (Odonata, Pseudostigmatidae) at a forest trail in Werehpai Camp. Both photographed by N. von Ellenrieder.

Plate V. a: Male of *Phyllogomphoides major* (Odonata, Gomphidae). **b**: Male of *Diastatops pullata* (Odonata, Libellulidae). Both photographed at Sipaliwini River, near Sipaliwini Camp, by N. von Ellenrieder.

Males of the clubtail dragonfly *Phyllogomphoides major* swiftly patrolled rivers and major creeks surrounding the Camps, and could be seen perching on tip of branches of low vegetation on the banks. Females only visit the water for oviposition, and larvae burrow on the muddy river beds.

Plate VI. a: Female of *Erythrodiplax castanea* (Odonata, Libellulidae) at a forest clearing in Sipaliwini Camp.
b: Male of *Fylgia amazonica lychnitina* (Odonata, Libellulidae) at a forest swamp in Werehpai Camp. **c**: Male of
Macrothemis hemichlora (Odonata, Libellulidae) at a forest stream in Sipaliwini Camp. All photographed by N. von
Ellenrieder.

Plate VII. a: Male of *Misagria parana* (Odonata, Libellulidae) at a forest clearing in Werehpai Camp.
b: Male of *Oligoclada walkeri* (Odonata, Libellulidae) at a forest stream in Sipaliwini Camp. Both photographed by N. von Ellenrieder.

Plate VIII. a: Male of *Perithemis cornelia* (Odonata, Libellulidae) at a forest stream in Sipaliwini Camp. **b**: Male of *Perithemis thais* (Odonata, Libellulidae) at a forest swamp in Werehpai Camp. **c**: Female of *Uracis oviposi-trix* (Odonata, Libellulidae) at a forest trail in Kutari Camp. All photographed by N. von Ellenrieder.

Dragonflies of the genus *Perithemis*, commonly known as 'Amber wings' due to their golden-orange wings, are believed to mimic wasps in their flight. Four species of this genus were encountered during this RAP; the pictures below show a male of *Perithemis cornelia* at a forest stream in Sipaliwini Camp and a male of *Perithemis thais* at a forest swamp in Werehpai Camp. Adults fly close over water in sunshine or in partially shaded areas. They perch on small branches or reeds usually close to water's surface, often with fore and hind wings slowly moving alternatively. Larvae crawl on substrate and among submerged vegetation.

Appendix A. List of Odonates from SW Suriname, Sipaliwini District, Kwamalasamutu region. Relative abundance: R (rare = 1–3 individuals seen); F (frequent = 4–20 individuals seen); C (common = 21–50 individuals seen). * = new record for Suriname; # = misidentified in the literature.

Family	Species	Site			
		Kutari	Sipaliwini	Werehpai	Kwamalasamutu/ Iwana Samu
		Relative abundance of species per site			
Calopterygidae (2 gen., 4 spp.)	*Hetaerina caja dominula* Hagen in Selys, 1853	C	C	C	F
	Hetaerina moribunda Hagen in Selys, 1853	R	F	F	-
	Hetaerina mortua Hagen in Selys, 1853	R	R	-	-
	Mnesarete cupraea (Selys, 1853)	R	-	-	-
Coenagrionidae (4 gen., 15 spp.)	*Acanthagrion ascendens* Calvert, 1909	-	R	F	F
	Acanthagrion chacoense Calvert, 1909	-	-	R	-
	Acanthagrion indefensum Williamson, 1916	R	-	-	-
	Acanthagrion rubrifrons Leonard, 1977	F	F	F	-
	Argia fumigata Hagen in Selys, 1865	R	-	-	-
	Argia insipida Hagen in Selys, 1865	F	R	-	-
	Argia oculata Hagen in Selys, 1865	F	C	F	C
	Argia translata Hagen in Selys, 1865	F	F	C	F
	* *Argia* sp. 1	R	C	C	F
	* *Argia* sp. 2	C	R	R	-
	* *Argia* sp. 3	F	-	-	-
	* *Argia* sp. 4	F	R	-	-
	# *Inpabasis rosea* (Selys, 1877)	F	F	F	-
	Metaleptobasis mauritia Williamson, 1915	-	-	R	-
	#*Metaleptobasis quadricornis* (Selys, 1877)	-	-	R	-
Dicteriadidae (1 gen., 1 sp.)	*Heliocharis amazona* Selys, 1853	R	R	R	-
Megapodagrionidae (2 gen., 3 spp.)	*Heteragrion ictericum* Williamson, 1919	F	R	F	-
	# *Heteragrion silvarum* Sjöstedt, 1918	R	R	R	-
	Oxystigma williamsoni Geijskes, 1976	-	-	F	-
Perilestidae (1 gen., 1 sp.)	*Perilestes attenuatus* Selys, 1886	-	-	R	-
	* *Perilestes gracillimus* Kennedy, 1941	R	R	F	-
	Perilestes solutus Williamson & Williamson, 1924	-	R	R	-
Protoneuridae (5 gen., 11 spp.)	*Epipleoneura fuscaenea* Williamson, 1915	R	F	F	-
	* *Epipleoneura pereirai* Machado, 1964	F	R	F	-
	* *Neoneura angelensis* Juillerat, 2007	R	-	R	-
	* *Neoneura denticulata* Williamson, 1917	R	R	R	-
	Neoneura joana Williamson, 1917	F	F	F	-
	Neoneura mariana Williamson, 1917	-	R	-	-
	Neoneura myrthea Williamson, 1917	F	F	F	R
	Phasmoneura exigua (Selys, 1886)	F	R	F	-
	Protoneura calverti Williamson, 1915	-	F	-	-
	Protoneura tenuis Selys, 1860	F	F	F	-
	Psaironeura tenuissima (Selys, 1886)	R	-	-	-
Pseudostigmatidae (2 gen., 3 spp.)	*Mecistogaster lucretia* (Drury, 1773)	R	F	F	F
	Mecistogaster ornata Rambur, 1842	R	-	-	-
	Microstigma anomalum Rambur, 1842	F	R	R	-

table continued on next page

Family	Species	Site			
		Kutari	Sipaliwini	Werehpai	Kwamalasamutu/ Iwana Samu
		Relative abundance of species per site			
Aeshnidae (3 gen., 6 spp.)	*Gynacantha auricularis* Martin, 1909	-	R	R	-
	Gynacantha gracilis (Burmeister, 1839)	-	-	R	-
	Gynacantha klagesi Williamson, 1923	-	-	R	-
	Gynacantha sp.	-	R	R	-
	Staurophlebia reticulata (Burmeister, 1839)	R	R	F	-
	Triacanthagyna ditzleri Williamson, 1923	-	R	-	-
Gomphidae (6 gen., 7 spp.)	*Aphylla* sp.	R	-	R	-
	Archaeogomphu nanus Needham, 1944	-	-	R	-
	Ebegomphus demerarae (Selys, 1894)	R	-	-	-
	Phyllocycla ophis (Selys, 1869)	-	R	-	-
	Phyllogomphoides major Belle, 1984	-	-	R	-
	Phyllogomphoides undulatus (Needham, 1944)	F	F	R	-
	Progomphus brachycnemis Needham, 1944	-	-	R	-
Libellulidae (19 gen., 40 spp.)	*Argyrothemis argentea* Ris, 1909	F	-	-	-
	Brechmorrhoga praedatrix Calvert, 1909	-	-	R	-
	Diastatops pullata (Burmeister, 1839)	R	F	R	-
	Dythemis multipunctata Kirby, 1894	-	R	F	-
	Elasmothemis cannacrioides (Calvert, 1906)	F	F	F	-
	* *Elasmothemis rufa* De Marmels, 2008	R	R	F	-
	Elga leptostyla Ris, 1909	-	R	-	-
	Erythrodiplax castanea (Burmeister, 1839)	R	-	-	-
	Erythrodiplax famula (Erichson, 1848)	R	F	F	C
	Erythrodiplax fusca (Rambur, 1842)	R	R	R	R
	Fylgia amazonica lychnitina De Marmels, 1989	R	-	R	-
	* *Gynothemis pumila* (Karsch, 1890)	R	-	-	-
	Macrothemis declivata Calvert, 1909	-	F	-	-
	Macrothemis delia Ris, 1913	R	-	-	-
	Macrothemis hemichlora (Burmeister, 1839)	-	F	-	-
	* *Macrothemis ludia* Belle, 1987	-	-	R	-
	Macrothemis sp.	R	-	-	-
	Miathyria simplex (Rambur, 1842)	-	-	-	C
	Micrathyria artemis Ris, 1911	-	-	-	R
	* *Micrathyria paruensis* Geijskes, 1963	R	-	R	-
	Micrathyria spinifera Calvert, 1909	R	R	-	-
	Misagria calverti Geijskes, 1951	-	-	R	-
	Misagria parana Kirby, 1889	-	-	F	-
	Nephepeltia flavifrons (Karsch, 1889)	-	-	-	F
	Oligoclada abbreviata (Rambur, 1842)	C	F	F	F
	Oligoclada amphinome (Ris, 1919)	F	-	F	-
	Oligoclada rhea (Ris, 1911)	-	-	-	R

table continued on next page

Family	Species	Site			
		Kutari	Sipaliwini	Werehpai	Kwamalasamutu/ Iwana Samu
		Relative abundance of species per site			
Libellulidae *(continued)* (19 gen., 40 spp.)	*Oligoclada walkeri* Geijskes, 1931	-	F	R	-
	Orthemis aequilibris Calvert, 1909	R	F	F	-
	* *Orthemis anthracina* De Marmels, 1989	-	-	R	-
	* *Orthemis coracina* von Ellenrieder, 2009	-	-	R	-
	Orthemis cultriformis Calvert, 1899	-	-	F	-
	Perithemis cornelia Ris, 1910	F	C	F	-
	Perithemis lais (Perty, 1834)	R	R	-	-
	Perithemis mooma Kirby, 1889	-	R	R	F
	Perithemis thais Kirby, 1889	C	C	C	F
	Tauriphila argo (Hagen, 1869)	-	-	-	F
	Uracis fastigiata (Burmeister, 1839)	-	-	F	F
	Uracis infumata (Burmeister, 1839)	F	-	R	-
	Uracis ovipositrix Calvert, 1909	F	R	R	-
	Uracis siemensi Kirby, 1897	-	-	R	-
10 families	**94 spp.; 45 genera**	**57 spp.; 31 genera**	**52 spp.; 28 genera**	**65 spp.; 34 genera**	**18 spp.; 13 genera**

Appendix B. Biology notes on odonates found during the Kwamalasamutu RAP survey.

Genus	Biology notes
Hetaerina	Adults perch on tips of branches and on leaves overhanging water of forest streams (Pl. I b, page 65) and rivers, where they wait for potential mates. More than one species usually occupies the same stream, or species assemblages partition different microhabitats (e.g. sunny versus shaded) of the same stream. Adults commonly found along forest trails perching on leaves and twigs on sunny spots (Pl. I a, page 65). Larvae live in submerged vegetation in aerated portion of streams.
Mnesarete	Adults perch on top of leaves overhanging water of forested streams, and larvae live among submerged aquatic vegetation.
Acanthagrion	Slow backwaters of forest streams and rivers (*A. ascendens*, Pl. II a [page 66], *A. chacoense, A. indefensum*), small pools in forest swamps (*A. rubrifrons,* Pl. II b, c [page 66]), and vegetated ditches and ponds in open areas (*A. ascendens*). Females oviposit inside stems of floating plants or masses of algae while in tandem with males, which often remain upright. Larvae live among submerged aquatic vegetation.
Argia	Some species inhabit forest rivers and streams (*A. fumigata, A. insipida, A. translata*), while others breed in forest swamps (*A. oculata, A.* sp. 1, *A.* sp. 2, *A.* sp 3. *A.* sp 4), where larvae live under stones and on sediment. Adults perch on rocks, logs, and twigs close to water surface, on overhanging leaves at streams, or on leaves, twigs and on the ground along forest trails near swamps (page 17), usually in sunny and bare substrates. After landing, adults open and close their wings two or three times. Females oviposit in masses of floating algae or aquatic plants, alone or while still in tandem, with males often standing vertically.
Inpabasis	Adults perch on leaves of bushes at sunny spots in forest swamps. Males where observed guarding little muddy depressions (page 17) where *I. rosea* is likely to breed. Larvae still unknown.
Metaleptobasis	Small sluggish streams and swampy areas within shaded forest zones; adults difficult to detect because they seldom fly and remain near ground, perching still on twigs and bushes. Larvae still unknown.
Heliocharis	Adults wary, perch on sunlight spots on vegetation overhanging water along forest streams with wings outspread. All adult odonates usually have spiny legs, which form a functional 'cage-net' to trap insects caught in flight. *Heliocharis* (and *Dicterias*) present an exception, as they have only minute spines on their legs, and presumably use their unusually large mouthparts instead to catch and hold their prey. Larvae inhabit stream pools with litter substrate and marginal areas of creeks and small slow-flowing rivers bordered by abundant riparian vegetation.
Heteragrion	Adults localized along shaded forest streams and rivers, perching by hanging with outspread wings on twig or leaf tips (Pl. III, page 67). Also commonly found on more open places along creeks and trails, where they catch prey from advantageous perches, usually returning to the same spots. Larvae live among leafy detritus and on lime and gravel beds of streams, and dwelling among organic matter and plant detritus where there is little or no water flow in primary and secondary forest streams.
Oxystigma	Forest streams, where adults perch on tips of twigs in the shade near creeks with wings outspread (Pl. III b, page 67), and can be active during heavy rain (Wasscher 1990), unique among odonates. Oviposition occurs in tandem, and males remain upright while holding females, which insert eggs into fallen palm leaves (Geijskes 1976). Emergence occurs on rocks along margins of creeks (De Marmels 1987).
Perilestes	Adults perch with partially open wings hanging vertically from vegetation in the shaded understory surrounding forest creeks (page 17), where they are rendered inconspicuous by their dark and dull colors. Female ovipositor is strong and likely used to lay eggs in hard substrates, including bark of twigs. Known larvae live among dead leaf litter accumulated in beds of quiet areas of small forest streams (Santos 1969, Neiss & Hamada 2010).
Epipleoneura	Small and dark inconspicuous adults hover very close to water surface and perch in shade under overhanging bushes along forest rivers. Tenerals perch on forest bushes in sunny spots away from water. Known larvae for other species of the genus live in rapidly flowing streams within primary forest (De Marmels 2007).
Neoneura	Adults hover just above water surface in forest streams and rivers, in bright or dappled sunshine, perching on twigs overhanging banks. Females oviposit in tandem in masses of small floating sticks, in dead grassy debris, sticks, and branches, or in muddy banks (Williamson 1918, De Marmels 1992, 2007). Larvae cling to stones in stream beds or are associated with vegetation or coarse detritus in river and creek pools (Westfall & May 2006).
Phasmoneura	Adults found in forest swamps, perching on tips of twigs and leaves (page 17). Larvae still unknown.
Protoneura	Small muddy or sandy streams within forest. Adults fly in shaded areas along banks, hover almost motionless near water surface, and perch on tip of small branches and twigs (Pl. IV a, page 68). They feed on small insects and some species are known to glean small spiders and flies from vegetation (Louton et al. 1996). Females oviposit alone or still in tandem, in leaves or pieces of wood floating in the water, debris, mosses, stems of aquatic plants, or roots, in areas of slow or standing waters, and several pairs can aggregate to oviposit.

table continued on next page

Genus	Biology notes
Psaironeura	Forest swamps, where adults fly among low vegetation in the shade, and perch on stems or leaf tips close to the ground. Larvae still unknown.
Mecistogaster	Adults perch on trees and bushes along forest trails (Pl. IV b, page 68) and clearings, and fly across forest understory in a slow wavy fashion, which combined with their unusually large size makes them very conspicuous. They glean spiders or wrapped spider prey items from spider webs, and females oviposit in species-specific phytotelmatha; larvae of *M. ornata* live in water accumulated in tree holes, other species breed in water filled tree holes (*M. jocaste*, *M. linearis*), bromeliad bracts (*M. modesta*), or bamboo holes (*M. asticta*, *M. jocaste*; Garrison et al. 2010). Larvae and breeding habitat of *M. lucretia*, the most common species in this study, are still unknown.
Microstigma	Appearance and habits as in *Mecistogaster*. Larvae live in water-filled tree holes and fallen fruit or nut husks (Neiss et al. 2008).
Gynacantha	Adults inconspicuous, due to their olive-brown ground color and crepuscular habits flying seldom through dense forest understory. They can be found perching on vegetation along margins of forest trails. Larvae inhabit mud bottomed pools, phytotelmatha, and streams (Bede et al. 2000).
Staurophlebia	Large size and bright contrasting green and red colors render adults conspicuous when they patrol forest rivers. Occasionally attracted to light at night. Larvae live in forest rivers (Geijskes 1959).
Triacanthagyna	Crepuscular habits in forests, where they breed in temporary pools and phytotelmatha. Adults can be found perching on vegetation along margins of forest trails.
Aphylla	Adults found along forest trails or at margins of streams where they alight on ground or snags with horizontally spread wings. Larvae are borrowers in soft muddy bottom zones in slowly flowing streams or still water areas (Belle 1992).
Archaeogomphus	Adults perch on overhanging leaves or twigs in dappled shade along forest stream banks, in which larvae live in pools with sand and detritus.
Ebegomphus	Forest rivers, where adults hover over sunny spots, low over the surface of quietly flowing water, and perch on twigs and dry branches. Larvae inhabit leafy trash in pools and eddies of creeks, and have been reported to emerge on large floating leaves at midday (Belle 1970).
Phyllocycla	Adults perch occasionally on dry twigs in the sun along forest rivers and streams, which males patrol flying swiftly close to water. Larvae probably inhabit silt-covered backwaters.
Phyllogomphoides	Males perch on stones or tip of branches of low vegetation on forest river banks (Pl. V a, page 69), and females visit the water only for oviposition. Larvae burrow on muddy river beds (Belle 1984).
Progomphus	Adults perch on rocks or low vegetation along forest river and stream banks. Larvae burrow in sand beds (Belle 1966b).
Argyrothemis	Adults found in forest swamps, where males are striking by flashing their bright pruinose-white thoracic dorsum when flying through a sun-patch, and disappearing as they move into the shade. Larvae camouflage themselves with detritus (Fleck 2003a).
Brechmorrhoga	Adults course up and down sections of forest rivers, and perch in pendent position in forest clearings or in brush on river banks. Larvae commonly found in areas of rocky substrate, shallow water, and rapid flow. The larva of *B. praedatrix* is apparently associated with the aquatic plant *Mourera fluviatilis* (Podostemaceae), which grows only in fast moving currents (Fleck 2004).
Diastatops	Marshy areas along forest rivers, where adults perch on vegetation along banks (Pl. V b, page 69), often with conspicuous black and red wings asymmetrically set, and larvae live among submerged vegetation.
Dythemis	Forest streams in sunny or shaded areas.
Elasmothemis	Forest rivers; adults perch on twigs with wings set. Females of *E. cannacrioides* deposit egg strands on floating roots of a liana (González-Soriano 1987).
Elga	Forests streams, where adults perch in sunny patches on leaves of trees bordering banks and fly down to water only for short periods at a time. Larvae live on stream bed.
Erythrodiplax	Forest stream pools and swamps, where larvae live. Adults perch with wings set on tips of grass, low stems, and twigs in open areas (Pl. V a, page 70).
Fylgia	Forest swamps; males land on leaves overhanging water in dappled sunlight, where their white face and red abdomen strike out in the mosaic of twilight and sunspot-shade (Pl. VI b, page 70). Larvae inhabit small, clear, stagnant pools with leaf litter (De Marmels 1992).
Gynothemis	Adults glide over clearings in little mini-swarms at head height and above, and perch on twigs on small creeks within forest where they breed.

table continued on next page

Genus	Biology notes
Macrothemis	Adults hover or course up and down sections of moderate to fast forest creeks, and often land on banks or vegetation bordering streams (Pl. VI c, page 70) or forest clearings; also found sometimes in swarms with one or more other species over glades in forest. Larvae burrow in areas of slow or still water, muddy substrate, and closed canopy.
Miathyria	Lentic environments with floating vegetation, among which roots the larvae live. Adults usually fly in swarms about 3 m over fields, in company with individuals of *Tauriphila*.
Micrathyria	Forest streams (*M. paruensis* and *M. spinifera*) and vegetated ditches (*M. artemis*). Males perch on tips of reeds and grass on shore vegetation with wings set. Females commonly found perching away from water on tips of twigs of bushes and trees.
Misagria	Adults perch on tips of branches and snags with wings set in partially shaded forest areas, and are commonly found in forest clearings and along trails (Pl. VII a, page 71). Larvae unknown.
Nephepeltia	Adults perch with wings set on tips of stems along marshy environments, in which they breed.
Oligoclada	Adults found along forest rivers (*O. abbreviata*), streams (*O. walkeri*, Pl. VII b, page 71), swamps (*O. amphinome*), and ditches (*O. rhea*) where they alight on ground or on surfaces of leaves. Known larva (*O. abbreviata*) collected in an artificial reservoir with strongly fluctuating water levels (Fleck 2003a).
Orthemis	Adults at slow moving forest streams, where males defend territories from a perch on reed, branch, or snag. Females oviposit guarded by the hovering males. Often also found perching in high sunny branches in forest clearings.
Perithemis	Forest streams (*P. cornelia, P. mooma, P. lais, P. thais*), swamps (*P. thais*), and vegetated ditches (*P. mooma*), where adults fly close over water in sunshine or in partially shaded areas. They perch on small branches or reeds usually close to water surface (Pls. VIII a, b, page 72) often with fore and hind wings slowly moving alternatively. Larvae crawl on substrate and among submerged vegetation.
Tauriphila	Adults engage in sustained gliding flights over open fields often in company with individuals of *Miathyria*; larvae live in temporary ponds and ditches among submerged vegetation.
Uracis	Adults perch on small twigs and vegetation near the ground in shaded forest understory or along partially shaded forest trails (Pl. VIII c, page 72). Females of *Uracis ovipositrix* lay eggs in damp earth and in the muddy bottom of small rain puddles. Males of *U. siemensi* hold territories at small water holes in the forest, close to creeks. Larvae unknown.

Chapter 4

Aquatic beetles of the Kwamalasamutu region, Suriname (Insecta: Coleoptera)

Andrew Short and Vanessa Kadosoe

SUMMARY

We conducted an intensive survey of aquatic beetles in the Kwamalasamutu Region of southwestern Suriname from 18 August to 7 September 2010. Both active collecting (using nets and by hand in aquatic habitats) and passive collecting (flight intercept traps, UV lights, dung baits) resulted in the collection of more than 4000 aquatic beetle specimens. We documented 144 species, distributed among 62 genera in 9 families. Sixteen of these species have been confirmed as new, with an additional 10 likely to be new. Two of these new species, both in the family Hydrophilidae, are described here: *Oocyclus trio* Short & Kadosoe sp.n. and *Tobochares sipaliwini* Short & Kadosoe sp.n. Camps 1 (Kutari) and 3 (Werehpai) had comparatively high species diversity, with 91 and 93 species respectively—although only 48 of these species were shared between the two sites. Camp 2 (Sipaliwini) had the lowest number of species with 68.

INTRODUCTION

Aquatic beetles represent a significant fraction of freshwater macroinvertebrate communities. At present, aquatic beetles are represented by nearly 13,000 described species distributed worldwide (Jäch & Balke 2008)—a guild richer in species than birds. These species are distributed across approximately 25 beetle families within four primary lineages: Myxophaga, Hydradephaga, aquatic Staphyliniformia (Hydrophiloidea & Hydraenidae) and the Dryopoidea (or aquatic Byrrhoids). Ecologically, these beetles play a variety of roles. Members of Myxophaga are small beetles that feed largely on algae as larvae and adults. The Hydradephaga (including the diving and whirligig beetles) are largely predators as adults and larvae; many aquatic Staphyliniformia are largely predators as larvae but scavengers as adults; the dryopoids are largely scavengers or eat algae as both larvae and adults.

Aquatic insects in general (including some groups of aquatic beetles) are often used to assess water quality in freshwater rivers and streams. The dryopoids are most frequently used for this purpose because they are most commonly found in these habitats and rely on highly oxygenated waters. Aquatic beetle communities are also effectively used to discriminate among different types of aquatic habitat (e.g. between lotic and lentic). However, in order to utilize aquatic insects as effective indicators of watershed health, these communities must be both 1) known and identifiable, and 2) have adequate information about their water quality tolerances. As neither of these criteria is met in the Guiana Shield region of South America, gaining more knowledge about both the diversity and ecology of these species is exceedingly

The RAP Bulletin of Biological Assessment meets all criteria for the description of species new to science, as specified by the International Commission on Zoological Nomenclature (ICZN). Paper copies of the RAP Bulletin are deposited in the library of Conservation International, Arlington, VA, USA; the Middleton Library at Louisiana State University, Baton Rouge, LA, USA; and the North Carolina Museum of Natural Sciences in Raleigh, NC, USA. The print run of this issue of the RAP Bulletin consisted of 500 copies. —BJO, LEA, THL, October 2011

helpful if macroinvertebrates are to play a role in water quality monitoring.

No prior surveys in Suriname have focused on aquatic beetles, and the fauna of the country, as well as the Guiana Shield region in general, remains very poorly known. Regionally, recent survey efforts of varying intensity in Venezuela, Guyana, and French Guiana have contributed to knowledge of the fauna (e.g. Short & Garcia 2011, Queney 2006, 2010, Short & Gustafson 2010a, b), but it is still a long way from being completed. In Venezuela alone, more than 50 new species have been described in the last five years, with at least twice that number of confirmed new species still awaiting description (Short, unpub. data). A preliminary review of the literature suggests approximately 75 aquatic beetle species are known from Suriname (Hansen 1999, Miller 2005, Nilsson 2001, Nilsson & van Vondel 2005, Short & Hebauer 2006, Short & Fikáček 2011, Ochs 1964, Spangler & Steiner 1983, Young 1971).

Here, we report on the findings of an intensive survey of aquatic beetles in the Kwamalasamutu region of southwestern Suriname, including the descriptions of two new species.

METHODS AND STUDY SITES

We collected aquatic beetles at all three main sites on the RAP (Site 1: Kutari; Site 2: Sipaliwini; Site 3: Werehpai). We also collected small, incidental samples at Iwana Samu.

Field methods
We employed a variety of passive and active collecting techniques to assemble as complete a picture of the aquatic beetle communities of the region as possible. Passive techniques are advantageous because they often allow large amounts of material to be collected in quantitative ways at one time and with little effort, but they provide little ecological or habitat data—and thus we do not gain new insights into the water quality requirements of insects collected in this manner. In contrast, active collecting methods (i.e. by hand) provide a richer source of information on the microhabitat and water quality requirements of species, but are more time intensive and qualitative, and may suffer from collector bias. Our survey focused on adult beetles, and although a few larvae were collected, they are not included in our results or analysis.

Traps and other passive methods. On most nights, we collected in the evening hours until approximately 10 p.m. at a UV light mounted on a white sheet erected on the periphery of each camp. We also used flight intercept traps (FITs) to sample the beetle fauna. These traps collect flying insects, including dispersing aquatic beetles. At Site 1 we used two FITs, each composed of a 2-meter wide by 1.5- meter high screen, with aluminum pans filled with soapy water as a collecting trough. At Sites 2 and 3, we used three FITs of slightly smaller dimensions. At all three camps, we examined the by-catch of dung traps set for the purpose of collecting Scarabaeinae, and set several dung traps of our own.

Active methods. For active collection of swimming insects, we used a large aquatic insect net to probe larger and deeper pools and river margins. We also targeted insects that float on the water's surface using small metal strainers to collect in micropools and marginal areas. We also collected in several 'niche' habitats, including the phytotelmata of *Heliconia* spp. at Site 3, the rock face seeps and damp soil on the inselberg at Site 2, and damp leaf litter at Site 3. For the latter, we submerged leaf packs in a tub of water and collected the insects that floated to the surface.

Site 1: Kutari. N 02°10′31″, W 056°47′14″, 200–250 m, 18–24 August 2010
Most collecting was conducted in streams and flooded forest that intersected the first 1.5 km of trails cut from camp. The majority of the terrain in the sampled area consisted of seasonally inundated forest, with a few patches of terra firme. Most sampled habitats included seasonally flooded forest, including muddy swamp-like areas, and flooded low-lying areas between hills that likely draw down to form streams. Most sampling sites were full of detritus. One stream in particular had substantially more sand/non-mud substrate than the others and was particularly rich in aquatic beetle taxa. The frequency of specimens arriving to the UV light was very uneven: most nights were relatively poor in diversity, but the evening of 20 August (when most of the light-trap diversity was collected) was a notable exception.

Habitats of note:
There were no particular habitat features of this site that were not present at the other sites.

Site 2: Sipaliwini. N 02°7′24″, W 056°36′26″, 200–250 m, 27 August–2 September 2010
All collecting was conducted at stream and swamp areas along a trail cut, approximately three kilometers in length, between camp and a small granite outcrop. Unlike the Kutari site, no expanses of flooded forest were observed. Several streams expanded slightly into adjacent forest, but only narrowly so as to give the appearance of a broader drainage rather than a swamp. Several larger ravines with more sand/gravel substrate were present in addition to the more typical detritus-filled streams. A stream that originated at the base of the granite outcrop and flowed around it was composed of both detritus and sandy substrate and was particularly high in beetle diversity.

Habitats of note:
There was a low, sloping granite outcrop (or inselberg) approximately three kilometers from the site. Parts of the outcrop were covered in scrubby vegetation. Algae and evaporation stains, as well as eroded grooves in the rock indicate seeps are present for part of the year. We were able to find one small seep with algae that had a very high density of beetles (Fig. 1 A-B).

*Site 3: Werehpai. N 02°21'47", W 056°1'52", 200–250 m,
3–7 September 2010*

Most collecting was conducted along an existing 3.5-km trail between the Sipaliwini River and the Werehpai caves. The camp area along the Sipaliwini was an abandoned farm site with secondary forest and an extremely dense understory, although this transitioned into more typical undisturbed forest relatively quickly along the trail. No flooded forest areas were observed at this site, although there was evidence that such areas exist during wetter times of the year (see habitats of note, below). Several small streams (<3 m) crossed the trail, some with detritus substrate and others with sand. Although the Werehpai cave area is an impressively large rock formation, it is not an inselberg, and no wet rock habitats or seepages were observed.

Habitats of note:

– About 1 km along the trail to the Werehpai caves, there is a c. 100-meter stretch where there were a few dozen water-filled depressions which varied from less than 1 m across to 5 m across, and a depth of very shallow to nearly 1 m. At higher water levels, the flooding of a

nearby stream might result in a flooded forest habitat in which these pools are connected. All of these pools were full of detritus and had incredible densities of aquatic beetles.

– Near the Werehpai caves, there was a small (2–3 meter wide), <1-meter deep meandering stream with a sandy substrate. The largely sand substrate distinguished this creek from most other streams we sampled during the expedition. A large number of very interesting species were found in this particular stream, and this may be due to the substrate type.

– Along part of the existing trail near the camp, a number of *Heliconia* plants were in bloom. We examined several of these inflorescences for beetles. We did not observe any *Heliconia* in bloom at the other sites.

Sample and specimen processing

Forty-nine samples were collected during the expedition, each representing a different stream, habitat, or trapping event. Most specimens were collected directly into 95% ethanol. Samples were subsequently mounted, labeled and databased. When long series of a single species were collected

Figure 1. Selected aquatic habitats. A-B) Sipaliwini site, inselberg habitat; C) Kutari site, sandy stream; D) Sipaliwini site, stream at base of inselberg.

from the same locality, only a portion of the specimens were mounted, and the remainder preserved in 80% ethanol. Total specimen counts given herein only reflect the mounted portion of the samples, so the abundance of the more common species is underreported.

RESULTS AND DISCUSSION

In total, 144 species of aquatic beetles in 62 genera were found, representing an extremely rich community, particularly given the relative homogeneity of the habitat examined (Appendix; Fig. 2). The Kutari and Werehpai sites had relatively high species diversity (91 and 93 species respectively), while the Sipaliwini site had only 68 species. Similarly, the Kutari and Werehpai sites had much higher numbers of site-unique species (32 and 23 respectively), while Sipaliwini had only seven (Table 1).

Of the 144 species recorded, 62 (43%) were found at only one of the three sites, whereas only 27 (18.7%) were found at all three. Based on both the number of shared species and Jaccard's index, the aquatic beetle communities found at the Sipaliwini and Werehpai sites were the most similar, while those found at Sipaliwini and Kutari the most dissimilar (Table 2). At the generic level, all three sites exhibited more comparable levels of diversity, with between 39 and 46 genera found per site.

We have confirmed that at least 16 of the species found are new to science (Appendix). One of these has already been described since the expedition (*Cetiocyon incantatus* Fikáček & Short 2010), and two are described here. We estimate that an additional 10–15 species are likely to be new to science, but additional research is needed to confirm their identities.

Not surprisingly, the fauna is typical of lowland Guiana Shield forest. Some taxa such as the genera *Siolius* Balfour-Browne, 1969, *Guyanobius* Spangler, 1986, *Fontidessus* Miller & Spangler, 2008, and *Globulosis* García, 2001 are

either endemic or largely restricted to the Guiana Shield. The fauna was very similar to what is known from southern Venezuela (south of the Orinoco River) and Guyana. The species found on and around the inselberg at the Sipaliwini site are restricted to this habitat type, and all likely have a very restricted range (perhaps endemic to the region and its periphery). Despite the presence of the inselberg at the Sipaliwini site, no species of Myxophaga were found. We suspect the isolation or lack of running water over rock contributed to the absence of this group.

Some taxa that we had initially expected to find were conspicuously absent. Not a single specimen of the common water scavenger beetle genus *Berosus* Leach, 1817 was found, despite having a high diversity in the region (more than 35 species are known from Venezuela, Oliva & Short, unpub. data). Many *Berosus* prefer the open waters of marshes and savannahs, and the absence of the genus during the RAP survey may be due to the vast unbroken forest that characterizes the Kwamalasamutu region. Other species that are often common and widespread in northern South America, such the ubiquitous *Tropisternus collaris* (Fabricius, 1775) and *T. lateralis* (Fabricius, 1775), also share these open-water habitats and were also not found during the survey.

Species of note

Cetiocyon incantatus Fikáček & Short, 2010: One specimen of this new species was collected from dung at Camp 1. Three specimens, also from Suriname, were already known from flight intercept traps, but this new record provided the first ecological information as to its habits. It is the first and currently only record for the genus *Cetiocyon* in the New World (Fikáček & Short, 2010).

Tropisternus phyllisae Spangler & Short, 2008: This species, only described three years ago, was known from a pair of specimens collected in 1962 along "Krakka-Phedra road", in the northern region of the country. A single specimen was collected in detrital pools at the Werehpai camp.

Tropisternus surinamensis Spangler & Short, 2008: This species, previously known from only a single female specimen, was described from the same locality as *T. phyllisae*. Specimens were taken together with *T. phyllisae* in this instance as well.

Hydrophilidae: "New genus 1": This genus, with two undescribed species, is in the tribe Acidocerini, and likely near the genera *Tobochares* or *Agraphydrus* Régimbart, 1903. It was found along the margins of sandy streams at both Sipaliwini and Werehpai camps. We are also aware of

Table 1. Aquatic beetle diversity by site.

	Specimens	Genera	Species	Site-unique species
Site 1/ Kutari	1501	46	91	32
Site 2/ Sipaliwini	1321	39	68	7
Site 3/ Werehpai	1586	43	93	23
Total	**4408**	**62**	**144**	-

Table 2. Species similarity between sites.

Sites	Shared Species	Non-Shared Species	Total Species	Jaccard's Index
Kutari-Sipaliwini	38	83	121	45.8%
Kutari-Werehpai	48	89	137	53.9%
Sipaliwini-Werehpai	50	69	119	72.5%

specimens of this genus from French Guiana (P. Queney, pers. comm.).

Hydrophilidae: "New genus 2": With a single undescribed species, this new genus is in the tribe Acidocerini, and likely near *Chasmogenus*. Like the other new acidocerine genus, it was taken at sandy streams at Sipaliwini and Werehpai camps. It is also known from French Guiana.

DESCRIPTION OF NEW SPECIES

We have chosen to describe two of the new species we discovered here, as the genera to which they belong already have been recently reviewed for the region. Thus, the species can be described and compared with existing species relatively easily. Some of the other lineages in which we

have confirmed new species are quite large, and will require significant further study to place them appropriately.

Genus *Oocyclus* Sharp, 1882

The genus *Oocyclus*, with nearly fifty described species, is found in both the Oriental and New World Tropics. Most species prefer hygropetric habitats, such as waterfalls and rock seepages. The first species from the Guiana Shield region were described last year (Short & Garcia, 2010).

Oocyclus trio Short & Kadosoe sp. n.

Type material: Holotype (male): "SURINAME: Sipaliwini District/ 2 10.973'N, 56 47.235'W, 210 m/ Camp 2, on Sipaliwini River/ leg. Short & Kadosoe; Inselberg/ 29–30. viii.2010; SR10-0829-01A/ 2010 CI-RAP Survey" (Deposited in the National Zoological Collection of Suriname).

Figure 2. Selected habitus images of aquatic beetle taxa collected in the Kwamalasamutu region. A) *Stegoelmis stictoides*, B) *Enochrus* sp. 2, C) *Pelonomus* sp. 1, D) *Copelatus geayi*, E) *Gyretes* sp. 1, F) *Vatellus tarsatus*, G) *Hydrochus* sp. 5, H) *Epimetopus* sp. 1, I) *Siolius* cf. *bicolor* (Images not to the same scale).

Paratypes (20): SURINAME: Sipaliwini District: Same data as holotype (13 exs.). Same camp but 1.ix.2010, seep on inselberg, SR10-0901-01A (7 exs.). Paratypes will be divided between the National Zoological Collection in Suriname, the University of Kansas, and the US National Museum of Natural History.

Diagnosis. The combination of the small body size, rounded posterolateral corners of the pronotum, lack of dense rows of setae on the elytra, a conspicuous circular white spot on the posterior third of each elytron, and dark brown abdominal ventrites with long setae will distinguish this species from other New World *Oocyclus*. When using the key of the Venezuelan species (Short & Garcia, 2010), one will arrive at couplet two, and then face choices that all clearly do not fit *O. trio*, as it has a different and distinctive suite of key characters that any of the other known species.

Description. *Size and form.* Body length = 3.7–4.0 mm. Body broadly oval, slightly convex (Fig. 3A). *Color.* Dorsum of head, pronotum, and elytra black, with a faint, irregular green sheen. Pronotal margins broadly pale along entire length except for posterolateral angles. Elytra with faint, indistinct green or bronze maculae; without a row of conspicuous rounded black [=without sheen] spots along suture. Each elytron with a small circular pale spot posteriorly. Maxillary and labial palps yellow; ventral face of head dark brown with stipes distinctly paler reddish brown. Lateral margins of prosternum and epipleura reddish brown, legs light brown to nearly yellow except for the dark brown coxae and basal margin of the femora. Abdominal ventrites dark brown. *Head.* Ground punctation on clypeus and frons moderately fine, distance between punctures 1.0–1.5× the width of one puncture. Systematic punctures on labrum consisting of some scattered indistinct punctures, the lateral

ones bearing short setae. Clypeus with a few very indistinct systematic punctures along anterolateral margins, slightly larger than surrounding punctuation. Frons with irregular row of systematic punctures mesad of each eye. Maxillary palps subequal to the width of labrum; palpomere 3 slight shorter than palpomere 2 in length, apical palpomere longer than penultimate. Labial palps about three-quarter as long as width of mentum. Mentum nearly smooth, with scattered moderately fine punctures; subquadrate, anterior margin slightly convex and depressed. *Thorax.* Ground punctation on pronotum and elytra evenly distributed and moderately coarse. Pronotal and elytral surface flat and even, without elevations or grooves. Pronotal systematic punctures with short fine setae, similar in size to ground punctures, mostly blending with larger ground punctures; anterior and posterior series each forming an irregular field. Posterolateral corners of pronotum rounded. Sutural punctation on elytra absent or unmodified from ground punctation; sutural interval not raised. Rows of systematic punctures on elytra present and moderately distinct, forming loose, rows of slightly larger punctures which may bear fine, short setae. Margins of elytra set with a few sparse setae, but not a dense fringe. Prosternum with a clearly defined median carina; slightly elevated anteromedially, the elevation set with 2 thickened spine-like setae. Elevated process of mesoventrite narrow and elongate, more than three times as long as wide, with 5 coarse spine-like setae. Metaventrite with narrow oval glabrous area posteromedially, ca. twice as long as wide, length of glabrous area ca. half the length of metaventrite. Procoxae with fine short pubescence and set with coarse, short spine-like setae. *Abdomen.* Ventrites with rather dense setae of varying lengths; each ventrite with scattered very long erect setae, distinctly longer than the longest setae on the metaventrite. Aedeagus as in Fig. 3B.

Habitat. All specimens were collected on and near a small inselberg. Specimens were found by disturbing the sand and detrital margins of a small stream that originated at the base of the inselberg, as well as on a small area of wet, algae covered rock on the inselberg itself.

Etymology. Named after the Trio people, the indigenous ethnic group of the region, who generously allowed us to conduct fieldwork and assisted with this survey on their land.

Genus *Tobochares* Short & Garcia, 2007

The genus *Tobochares* was described to accommodate a single species, *T. sulcatus* Short & Garcia, 2007, found along the northwestern edge of the Guiana Shield in the Venezuelan state of Amazonas. The species described here fits the generic diagnosis of *Tobochares* extremely well, and no modifications to the generic description are necessary with the inclusion of *T. sipaliwini*. *Tobochares sulcatus* generally occurs along the margins of streams and small rivers with granite bedrock with detritus. While the collecting events for the new species here described did not occur in streams flowing over granite, they were found in close proximity to an inselberg.

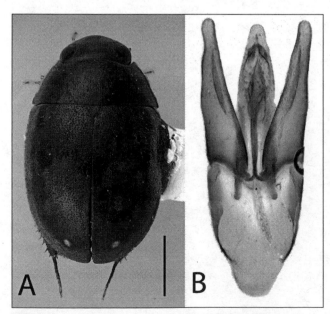

Figure 3. *Oocyclus trio* Short & Kadosoe sp. n. A) Dorsal habitus, scale bar = 1.0 mm, B) Aedeagus.

Tobochares sipaliwini Short & Kadosoe sp. n.

Type material: Holotype (male): "SURINAME: Sipaliwini District/ 2 10.973'N, 56 47.235'W, 210 m/ Camp 2, on Sipaliwini River/ leg. Short & Kadosoe; Inselberg/ 29–30. viii.2010; SR10-0829-01A/ 2010 CI-RAP Survey" (Deposited in the National Zoological Collection of Suriname). **Paratypes (4): SURINAME: Sipaliwini District:** Same data as holotype (3 exs.). Same camp but 31.viii.2010, sandy forest creek, SR10-0831-01B (1 ex.). Paratypes will be divided between the National Zoological Collection in Suriname, the University of Kansas, and the US National Museum of Natural History.

Diagnosis. Elytra sulcate on basal half only on disc (elytral sulcate on entire length in *T. sulcatus*, with medial series reaching the elytral base). Maxillary palps uniformly pale (apex darkened in *T. sulcatus*). Process of the mesoventrite distinctly elevated into a sharp tooth (process not elevated or acute in *T. sulcatus*). Outer margins of parameres straight (sinuate in *T. sulcatus*).

Description. *Size and form.* Body length = 1.8–2.0 mm. Body elongate oval (Fig. 4A), moderately dorsoventrally compressed. *Color and punctation.* Dorsum of head, pronotum and elytra very dark brown with the lateral margins of pronotum and elytra slightly paler. Anterolateral margins of clypeus with very faint paler preocular patches. Meso- and metathoracic ventrites and visible abdominal sterna dark brown, with prosternum, epipleura, and legs distinctly paler. Ground punctation on head, pronotum and elytra moderately fine. *Head.* Antennae with scape and pedicel subequal in length, and their combined length subequal to antennomeres 3–8. Maxillary palps with palpomeres 2 and 4 subequal in length with palpomere 3 slightly shorter. *Thorax.* Elytra with ten rows of serial punctures which are depressed into grooves in posterior half on the mesal region, to posterior four-fifths on the lateral region. Elevation of

mesoventrite forming a lateral carina which is raised into an acute tooth elevated to the same plane as the ventral surface of mesocoxae. Metaventrite with distinct median ovoid glabrous area that is nearly two-thirds as long as the metaventrite length. *Abdomen.* Abdominal ventrites uniformly and very densely pubescent. Apex of fifth ventrite evenly rounded. Aedeagus with basal piece short, ca. one-third as long as parameres. Parameres with inner and outer margins straight in apical half. Dorsal strut slightly extended past the apex of parameres, with gonopore of median lobe situated distinctly below apex of the dorsal strut.

Habitat. All specimens were collected in small creeks (<2 m wide) with a predominately sand/gravel substrate and with substantial amounts of detritus. The larger series was taken at a stream that originates at the base of a small inselberg.

Etymology. Named after the Sipaliwini River and region, near and in which this new species was found.

CONSERVATION ISSUES AND RECOMMENDATION

The relatively rich water beetle diversity was expected given the diverse complement of aquatic habitats available at each camp. The relatively high number of genera and species, which cover a variety of ecological and habitat types, suggest the area is largely undisturbed. Differences between the communities found at each camp seem likely due to either the presence of rare species ('singletons') or habitats found at some but not all camps. For example, only a few specimens of *Vatellus tarsatus* (LaPorte, 1835) were found at Site 2 in detrital pools, but as this species is very rare and probably patchy in its distribution, our failure to detect it at Sites 1 and 3 does not reflect poorly on those sites. Similarly, the genera *Oocyclus, Fontidessus*, and *Platynectes* Régimbart, 1878 were only collected at Site 2, but these taxa are usually restricted to rock outcrops. Since no outcrops were present in the vicinities of Sites 1 and 3, these species were not found. No substantive differences in the water beetle communities between the sites could be attributed to anthropogenic disturbances. Indeed, the lack of several common groups that typically prefer open habitats (e.g. *Berosus, Tropisternus collaris, T. lateralis*) — all of which are widespread in South America and known from Suriname — supports the observation of an intact forest landscape.

It is clear that a tremendous amount of basic survey and taxonomy work is needed to gain a complete picture of the aquatic beetle fauna of southern Suriname, as well as the Guiana Shield in general. Additional work in the region should take into account potential variations in seasonality and habitat composition, as the abundance and nature of flooded forest and stream margin habitats changes dramatically over the course of a year. Additional surveys in moderately disturbed or open-canopy sites would clarify whether the lack of certain taxa is due to habitat alteration or other artifacts in their native distributions.

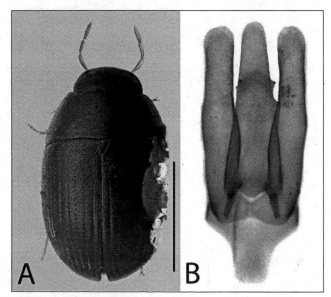

Figure 4. *Tobochares sipaliwini* Short & Kadosoe sp. n. A) Dorsal habitus, scale bar = 1.0 mm, B) Aedeagus.

REFERENCES

Fikáček, M. & A.E.Z. Short. 2010. Discovery of the genus *Cetiocyon* in the Neotropical Region and revision of its New Guinean species (Coleoptera: Hydrophilidae: Sphaeridiinae). *Arthropod Systematics and Phylogeny* 68(3): 309–329.

Gustafson, G.T. & A.E.Z. Short. 2010a. Redescription of the Neotropical water scavenger beetle genus *Phaenostoma* (Coleoptera: Hydrophilidae) with description of two new species. *Acta Entomologica Musei Nationalis Pragae* 50(2): 459–467.

Gustafson, G.T. & A.E.Z. Short. 2010b. Revision of the Neotropical water scavenger beetle genus *Guyanobius* Spangler (Coleoptera: Hydrophilidae: Chaetarthriini). *Aquatic Insects* 32(4): 245–258

Hansen, M. 1999. Hydrophiloidea (s. str.) (Coleoptera). In: World Catalogue of Insects, vol. 2. Stenstrup, Denmark, 416 pp.

Jäch, M. & M. Balke. 2008. Global diversity of water beetles (Coleoptera) in freshwater. Hydrobiologia 595: 419–442.

Miller, K.B. 2005. Revision of the New World and Southeast Asian Vatellini (Coleoptera: Dytiscidae: Hydroporinae) and phylogenetic analysis of the tribe. Zoological Journal of the Linnean Society, 144: 415–510.

Nilsson, A.N. 2001. Dytiscidae (Coleoptera). In: World catalogue of insects, vol. 3. Stenstrup, Denmark, 395 pp.

Nilsson, A.N. & van Vonel, B.J. 2005. Amphizoidae, Aspidytidae, Haliplidae, Noteridae, and Paelobiidae (Coleoptera). In: World Catalogue of Insects, vol. 7. Stenstrup, Denmark, 170 pp.

Ochs, G. 1964. Zur Kenntnis der Gyriniden (Col.) von Suriname und vom Rio Parú im benachbarten Brasilien. *Uitg. natuurw Studkring Suriname,* 27: 82–110.

Queney, P. 2006. Description of three new species of *Berosus* from French Guiana (Coleoptera, Hydrophilidae). *Bulletin de la Société Entomologique de France* 111: 457–464.

Queney, P. 2010. Three new species of *Notionotus* Spangler from French Guiana and Guyana (Coleoptera: Hydrophilidae). Koleopterologische Rundschau 80: 129–137.

Short, A.E.Z. & M. García. 2007. *Tobochares sulcatus*, a new genus and species of water scavenger beetle from Amazonas State, Venezuela (Coleoptera: Hydrophilidae). *Aquatic Insects* 29: 1–7.

Short, A.E.Z. & Hebauer, F. 2006. World Catalogue of Hydrophiloidea – additions and corrections, 1 (1999–2005) (Coleoptera). Koleopterologische Rundschau 76: 315–359.

Short, A.E.Z. & Fikáček, M. 2011. World catalogue of the Hydrophiloidea (Coleoptera): additions and corrections II (2006–2010). Acta Entomologica Musei Nationalis Pragae, 51.

Spangler, P.J. & A.E.Z. Short. 2008. Three new species of Neotropical *Tropisternus* Solier (Coleoptera: Hydrophilidae). *Zootaxa* 1917: 65–68.

Spangler, P.J. and Steiner, W.E. 1983. New Species of Water Beetles of the Genera *Elmoparnus* and *Pheneps* from Suriname (Coleoptera: Dryopidae; Psephenidae). Proceedings of the Entomological Society of Washington, 85, 826–839.

Young, F.N. 1971. Two new species of *Hydrodessus* from Suriname, with a key to the known species (Coleoptera: Dytiscidae). Uitgaven Natuurw Studkring Suriname, (61): 152–158.

Appendix. List of aquatic beetles collected on the Kwamalasamutu RAP survey. *=confirmed new species.

Taxon	Total # specimens	Kutari	Sipaliwini	Werehpai
DYTISCIDAE				
Anodocheilus sp. 1	24	X	X	X
Bidessodes sp. 1	15	X		
Bidessodes sp. 2	4	X		
Bidessodes sp. X	115	X	X	
Bidessonotus sp. 1	23		X	X
Celina sp. 1	5	X	X	
Copelatus geayi Régimbart, 1904	43	X	X	X
Copelatus sp. 2	14	X	X	X
Copelatus sp. 3	31	X	X	X
Copelatus sp. 4	16		X	X
Copelatus sp. 5	184	X	X	X
Copelatus sp. 6	125	X	X	X
Copelatus sp. 7	7	X	X	X
Copelatus sp. 8	2		X	X
Copelatus sp. 9	123	X	X	X
Copelatus sp. 10	1	X		
Derovatellus sp. 1	3	X	X	
Desmopachria sp. 1	16	X		
Desmopachria sp. 2	6	X	X	X
Desmopachria sp. 3	15	X	X	X
Desmopachria sp. 4	3	X		X
Desmopachria sp. 5	3	X		
Desmopachria sp. 6	7		X	X
Desmopachria sp. X	188	X	X	X
Fontidessus sp. 1*	128		X	
Hemibidessus sp. 1	26		X	X
Hemibidessus sp. 2	39	X	X	
Hydaticus subfasciatus Laporte, 1835	28	X	X	X
Hydaticus sp. 2	2		X	X
Hydrodessus sp. 1	1	X		
Hydrodessus sp. 2	1	X		
Hydrodessus sp. 3	1	X		
Hypodessus sp. 1	55	X		
Hypodessus sp. 2	14	X	X	
Laccodytes apalodes Guignot, 1955	33	X		
Laccodytes sp. 2	2	X		
Laccodytes sp. 3	1	X		
Laccodytes sp. 4	2	X		
Laccodytes sp. 5	2	X		X
Laccophilus sp. 1	8		X	X
Laccophilus sp. 2	28		X	X
Laccophilus sp. 3	11	X		X

table continued on next page

Taxon	Total # specimens	Kutari	Sipaliwini	Werehpai
Laccophilus sp. 4	1	X		
Laccophilus sp. 5	4	X		X
Laccophilus sp. 6	72	X	X	X
Laccophilus sp. 7	18		X	X
Laccophilus sp. 8	101	X	X	X
Laccophilus sp. 9	97	X	X	X
Pachydrus sp. 1	7		X	X
Platynectes sp. 1	50	X		
Thermonectus sp. 1	7			X
Thermonectus sp. 2	7		X	X
Thermonectus sp. 3	43			X
Thermonectus sp. 4	7			X
Uvarus sp. 1	20	X		
Vatellus tarsatus (LaPorte, 1835)	13			X
Vatellus sp. 2	2	X	X	
DRYOPIDAE				
Dryops sp. 1	29			X
Pelonomus sp. 1	20	X	X	
ELMIDAE				
Cyllepus sp. 1	1			X
Hexacylloepus sp. 1	4	X		X
Hintonelmis sp. 1	2		X	
Hintonelmis sp. 2	3	X		X
Neoelmis sp. 1	57	X	X	X
Neoelmis sp. 2	2	X		
Neoelmis sp. 3	17	X		X
Pagelmis sp. 1	1	X		
Pagelmis sp. 2	4		X	X
Pagelmis sp. 3	1			X
Pilielmis sp. 1	2	X		
Stegoelmis stictoides Spangler, 1990	21	X	X	X
Stenhelmoides sp. 1	65	X	X	
New Genus 1, sp. 1*	1	X		
New Genus 1, sp. 2*	32	X		
EPIMETOPIDAE				
Epimetopus sp. 1	5			X
Epimetopus sp. 2	2			X
GYRINIDAE				
Gyretes sp. 1	27	X		X
Gyretes sp. 2	31	X		X
Gyretes sp. 3	82	X		X
Gyretes sp. 4	11	X	X	
Gyretes sp. 5	2	X		

table continued on next page

Taxon	Total # specimens	Kutari	Sipaliwini	Werehpai
HYDRAENIDAE				
Hydraena paeminosa Perkins, 1980	13		X	X
Hydraena sp. 1*	1			X
Hydraena sp. 2*	1			X
HYDROCHIDAE				
Hydrochus sp. 1	2	X		
Hydrochus sp. 2	6			X
Hydrochus sp. 3	3		X	X
Hydrochus sp. 4	1		X	
Hydrochus sp. 5	2	X		
HYDROPHILIDAE				
Anacaena cf. *suturalis*	152	X	X	X
Australocyon sp. 1	6			X
Cercyon sp. 1	9	X		X
Cercyon sp. 2	17	X		X
Cetiocyon incantatus Fikacek & Short, 2010*	1	X		
Chaetarthria sp. 1	1			X
Chasmogenus sp. X*	248	X	X	X
Derallus intermedius Oliva, 1995	22	X	X	X
Derallus sp. 1	46	X		X
Derallus sp. 2	53	X		X
Derallus sp. 3	25		X	
Derallus sp. 4	2			X
Enochrus sp. 1*	189	X	X	X
Enochrus sp. 2	52		X	
Enochrus sp. 3	92		X	X
Enochrus sp. 4	3		X	X
Enochrus sp. 5	15		X	X
Enochrus sp. 6	123	X	X	X
Globulosis sp. 1	11	X		X
Guyanobius sp. 1	3	X		
Helochares sp. 1*	30	X	X	X
Helochares sp. 2	30	X	X	X
Helochares sp. 3	10	X		X
Helochares sp. 4*	10			X
Helochares sp. 5	1	X		
Hydrobiomorpha sp. 1	2			X
Hydrobiomorpha sp. 2	1	X		
Hydrophilus smaragdinus Brullé, 1837	5			X
Moraphilus sp. 1	7	X	X	
Notionotus shorti Queney, 2010	41		X	X
Oosternum sp. 1	1	X		
Oosternum sp. X	16	X	X	

table continued on next page

Taxon	Total # specimens	Kutari	Sipaliwini	Werehpai
Oocyclus trio Short & Kadosoe, sp.n.*	22		X	
Pelosoma sp. 1	14			X
Pelosoma sp. 2	6	X		
Pelosoma sp. 3	1			X
Pemelus sp.*	1	X		
Phaenonotum sp. 1	16		X	X
Phaenonotum sp. 2	3	X		
Phaenostoma sp. 1	4	X		X
Phaenostoma sp. 2	14			X
Phaenostoma sp. 3	38	X		X
Phaenostoma sp. 4	12	X		
Tobochares sipaliwini Short & Kadosoe, sp.n.*	6		X	
Tropisternus chalybeus	89	X	X	X
Tropisternus phyllisae Spangler & Short, 2008	1			X
Tropisternus setiger	13		X	X
Tropisternus surinamensis Spangler & Short, 2008	23		X	X
New Genus 1, sp. 1*	50	X		X
New Genus 1, sp. 2*	29	X		X
New Genus 2, sp. 1*	4			X
NOTERIDAE				
Notomicrus sp 1.	121		X	X
Notomicrus sp. X	239	X	X	X
Siolius cf. *bicolor*	56	X		X
Suphisellus sp. 1	168	X	X	X
TOTAL:	**4409**	**91**	**68**	**93**
Site-Unique Species:		**32**	**7**	**23**

Chapter 5

Dung beetles of the Kwamalasamutu region, Suriname (Coleoptera: Scarabaeidae: Scarabaeinae)

Trond H. Larsen

SUMMARY

Dung beetles are among the most cost-effective of all animal taxa for assessing biodiversity patterns, but relatively little is known about the dung beetle fauna of Suriname. I sampled dung beetles using baited pitfall traps and flight intercept traps in the Kwamalasamutu Region of southern Suriname. I collected 4,554 individuals represented by 94 species. Species composition and abundance varied quite strongly among sites. Dung beetle diversity correlated positively with large mammal species richness, and was highest at the most isolated site (Kutari), suggesting a possible cascading influence of hunting on dung beetles. Small-scale habitat disturbance also caused local dung beetle extinctions.

The dung beetle fauna of the Kwamala region is very rich relative to other lowland forests of Suriname and the Guianas, and contains a mix of range restricted endemics, Guiana Shield endemics, and Amazonian species. I estimate that about 10–15% of the dung beetle species collected here are undescribed. While most species were coprophagous, 26 species were never attracted to dung; 4 of these were attracted exclusively to carrion or dead invertebrates and the other 22 were only captured in flight intercept traps. The abundance of several large-bodied dung beetle species in the region is indicative of the intact wilderness that remains. These species support healthy ecosystems through seed dispersal, parasite regulation and other processes. Maintaining continuous primary forest and regulating hunting (such as through hunting-restricted reserves) in the region will be essential for conserving dung beetle communities and the ecological processes they sustain.

INTRODUCTION

Dung beetles (Coleoptera: Scarabaeidae: Scarabaeinae) are an ecologically important group of insects. By burying dung as a food and nesting resource, dung beetles contribute to several ecological processes and ecosystem services that include: reduction of parasite infections of mammals, including people; secondary dispersal of seeds and increased plant recruitment; recycling of nutrients into the soil; and decomposition of dung as well as carrion, fruit and fungus (Nichols et al. 2008). Dung beetles are among the most cost-effective of all animal taxa for assessing and monitoring biodiversity (Gardner et al. 2008a), and consequently are frequently used as a model group for understanding general biodiversity trends (Spector 2006). Dung beetles show high habitat specificity and respond rapidly to environmental change. Since dung beetles primarily depend on dung from large mammals, they are excellent indicators of mammal biomass and hunting intensity. Dung beetle community structure and abundance can be rapidly measured using standardized transects of baited traps, facilitating quantitative comparisons among sites and studies (Larsen and Forsyth 2005).

METHODS

I sampled dung beetles at all three sites (Kutari, Sipaliwini, and Werehpai) using standardized pitfall trap transects. Ten traps baited with human dung were placed 150 m apart along a linear transect at each site (see Larsen and Forsyth 2005 for more details). Traps consisted of 16 oz plastic cups buried in the ground and filled with water with a small amount of liquid detergent. A bait wrapped in nylon tulle was suspended above the cup from a stick and covered with a large leaf. At each site, traps were collected every 24 hours for four days, and were re-baited after two days. I set three flight intercept traps at each site to passively collect dung beetle species that are not attracted to dung. I also placed additional pitfall traps whenever possible with other types of baits that included rotting fungus, carrion, dead millipedes, and injured millipedes. All traps were collected daily. I opportunistically collected dung beetles that I encountered in the forest, usually perching on leaves during both day and night.

From August 19–24, 2010, I collected dung beetles at the Kutari site (N 02° 10' 31", W 056° 47' 14") in primary forest characterized by small hills and several swampy areas. From August 27 – September 4, 2010, I collected dung beetles at the Sipaliwini site (N 02° 17' 24", W 056° 36' 26") in primary forest with small hills and relatively dry, hard soils with high bedrock. From September 2–7, 2010, I collected dung beetles at the Werehpai site (N 02° 21' 47", W 056° 41' 52") in primary forest as well as in bamboo (1 dung trap) and secondary forest (1 dung trap). Beetles were identified and counted as they were collected in the field and voucher specimens were stored in ethanol for further study and museum collections. Beetle specimens are deposited at the National Museum of Natural History at the Smithsonian Institution in Washington, DC, USA and at the National Zoological Collection of Suriname in Paramaribo.

To estimate total species richness at each site and assess sampling completeness, I compared the observed number of species with the expected number of species on the basis of randomized species accumulation curves computed in EstimateS (version 7, R. K. Colwell, http://purl.oclc.org/ estimates) (Colwell and Coddington 1994). I used an abundance-based coverage estimator (ACE) because it accounts for species abundance as well as incidence, providing more detailed estimates. I also used EstimateS to calculate similarity among sites, using the Morisita-Horn similarity index which incorporates species abundance as well as incidence.

RESULTS AND DISCUSSION

I sampled a total of 94 species and 4,554 individuals of dung beetles during the RAP (Table 1, Appendix A). Species richness was similar at all sites. Among dung traps, for which sampling effort was identical at all sites, species richness was highest at Sipaliwini (49 species), followed by Werehpai

(47 species) and Kutari (44 species) (Table 1, Fig. 1). Species accumulation curves for dung-baited pitfall traps (based on abundance-based coverage estimator) indicated that I sampled an estimated 88% of all coprophagous species occurring in the area. However, sampling completeness was lowest at Werehpai where I sampled only 72% of the dung-feeding species likely to occur at the site (Table 1, Fig. 1). Consequently, species richness estimators predict that Werehpai supports the highest number of coprophagous species (65 species), followed by Kutari (60 species) and Sipaliwini (57 species) (Table 1).

These differences between observed and predicted species richness are probably explained by strong differences in abundance among sites. As with observed species richness, abundance was highest at Sipaliwini and lowest at Kutari; Sipaliwini supported almost three times as many individuals as Kutari (Table 1, Fig. 1). Low abundance at Kutari may have been influenced by the large areas of swamp and

Table 1. Diversity and abundance of dung beetles in Kwamala region.

	All sites	Kutari	Sipaliwini	Werehpai
Species richness (all samples)	94	70	62	67
Species richness (dung traps)	68	45	49	47
Estimated richness (ACE) (dung traps)	77	60	57	65
% Sampling completeness (dung traps)	88	75	86	72
Shannon diversity (H) (dung traps)	2.84	2.85	2.57	2.78
Abundance/trap (all samples)	23.6	13.8	34.3	23.5
Abundance/trap (dung traps)	33.9	16.7	49.3	35.7

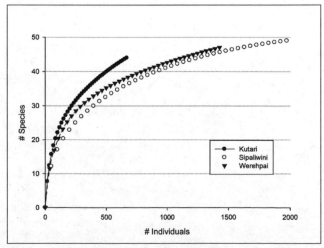

Figure 1. Species accumulation curves for each site based on dung-baited pitfall traps (40 trap samples for each site).

flooded forest at the site, conditions which negatively affect many dung beetle species whose larvae develop in the soil. However, diversity, measured by the Shannon index, showed the opposite pattern to observed species richness. Diversity was highest at Kutari and lowest at Sipaliwini, due to greater evenness of species' abundance distributions at Kutari (Table 1).

Out of 94 species sampled during this RAP survey, only 68 were attracted to dung. Considering all trap types and capture methods, Kutari supported the greatest number of dung beetle species (70 species) (Appendix A, B). Four species were attracted only to carrion or to dead invertebrates (Appendix B). 22 species were sampled only in flight intercept traps (Appendix B), and many of these species are poorly represented in collections because they are difficult to sample and in some cases, their diet is unknown. Some of these species show unusual specializations, such as millipede predation or colonization of leaf-cutter ant nests (see interesting species discussion below).

Species composition and community structure varied strongly among sites (Table 2). Sipaliwini and Werehpai were relatively similar in terms of community structure, showing a high Morisita-Horn index. Kutari was very distinct from both Sipaliwini and Werehpai, and contained many species not present at the other sites. Some of the most abundant species at a particular site were rare or completely absent from other sites (Appendix A). For example, I caught 211 individuals of *Ateuchus simplex* at Sipaliwini, and none at Kutari, despite the relative close proximity of both sites.

Dung beetle species richness was strongly reduced by habitat disturbance. Second growth forest supported only 70% of the total species richness found in primary forest, while bamboo supported only 40% of primary forest species richness (Fig. 2). Only one species, *Uroxys gorgon*, occurred in bamboo or secondary forest but did not occur in primary forest. *Uroxys gorgon* is known to be phoretic in sloth fur, and sloths are often hyper-abundant in secondary forest. The absence of other disturbance-adapted species in the Kwamala region was somewhat surprising, given the high number of 'weedy' species found in other parts of South America. Their absence might be explained by the extraordinarily low proportion of disturbed habitats occurring in southern Suriname.

Dung beetle diversity (measured by the Shannon index) was strongly positively correlated with species richness of

large mammals (Fig. 3), with the highest beetle diversity and mammal richness occurring at Kutari. Kutari also appeared to support the most primate species of all sites (see Large Mammals Chapter), and primates provide one of the most important food sources for dung beetles. High dung beetle diversity at Kutari may have been influenced by higher mammal richness and by lower hunting intensity, although further data are needed. On the other hand, dung beetle species richness and abundance were not correlated with the large mammal community, although no robust analysis was possible due to the short sampling period for mammals and the small number of sites for both groups (N=3). Furthermore, dung beetle abundance and species richness may have been influenced by differences in habitat and soil conditions, as discussed above.

At least 23 dung beetle species sampled during this RAP survey are known to be distributed across the Amazon basin. Many of the Amazonian species were locally rare and sampled at Kutari (Appendix A), which was the southernmost site sampled during the RAP. Out of these 23 Amazonian species, 20 occurred at Kutari, and only 16 at Werehpai and 15 at Sipaliwini. The Kwamala area may straddle the northern range limit for these species.

For the few genera that have been revised and for which good distributional data exist, many of the remaining species are restricted to the northern Amazon region, the Guiana Shield, or show an even more restricted range, while several are data deficient (see also interesting species discussion below). For example, *Coprophanaeus parvulus*, *Oxysternon festivum*, and *Eurysternus balachowskyi* are endemic to the Guiana Shield and northern Amazon, while *Oxysternon durantoni* and *Eurysternus cambeforti* occur only in the extreme northeastern Guianas (Edmonds and Zidek 2004, Genier 2009, Edmonds and Zidek 2010).

Dung beetle species richness is high in the Kwamala region relative to other areas in northeastern South America and the Guianas (Table 3). Similar RAP surveys at Lely and Nassau in Suriname yielded only 35–48% of the species

Table 2. Dung beetle community similarity among sites.

1st	2nd	S 1st	S 2nd	Shared Species	Morisita-Horn
Kutari	Sipaliwini	44	49	35	0.57
Kutari	Werehpai	44	47	30	0.61
Sipaliwini	Werehpai	49	47	37	0.90

N = 40 dung traps at each site, S = species richness

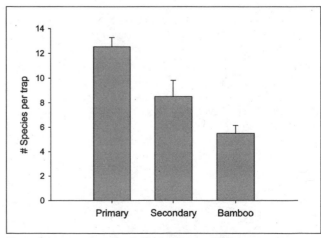

Figure 2. Impacts of habitat disturbance on species richness: primary forest, secondary forest and bamboo (mean ±1 SE).

richness found around Kwamala. Other studies from Venezuela, French Guiana, and Brazil also showed lower species richness in lowland primary forest with comparable sampling effort. Further sampling around Kwamala may yield as many or more species than were found in French Guiana and at Jari, Brazil, where greater sampling effort was employed (Table 3).

Table 3. Comparison of dung beetle species richness in primary lowland forests in northeastern South America.

	Kwamala region	Nassau, Suriname[1]	Lely, Suriname[1]	Guri, Venezuela[2]	Nouragues, F. Guiana[3,4]	Kaw Mtn, F. Guiana[4]	Jari, Amapa , Brazil[5]	Marajoara, Para, Brazil[6]
S (all samples)	94	27	38	41				
S (dung traps)	68	24	33	24 (32)	42 (78)	33 (47)	41–51 (72)	47

First number indicates species richness observed with comparable sampling effort to this RAP survey. Number in parentheses indicates species richness observed with more extensive long-term sampling effort, or across a broader landscape. [1]Larsen 2007; [2]Larsen et al. 2008, Larsen unpub. data; [3]Feer 2000; [4]Price & Feer in prep.; [5]Gardner et al. 2008b; [6]Scheffler 2005

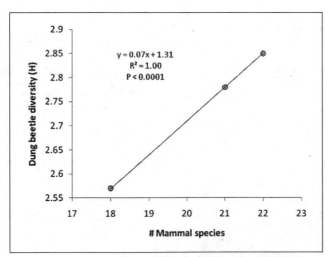

Figure 3. Linear regression of dung beetle diversity (Shannon diversity index) against large mammal species richness across all three sites.

INTERESTING SPECIES

I estimate that about 10–15% of the dung beetle species collected during this RAP (10 to 14 species) are undescribed. However, most of the genera collected here have never been revised, and determination of these undescribed species will require further comparisons with other museum collections. I sampled 26 species of *Canthidium* in the Kwamala area. *Canthidium* is a hyper-diverse yet very poorly known genus, and many of these species are almost certainly new to science. *Ateuchus* is also a poorly known yet diverse genus, and several *Ateuchus* species from the RAP are likely to be new. *Canthon* sp. 2 represents an undescribed species that is currently under study (see Appendix A).

Several large-bodied dung beetle species, such as *Coprophanaeus lancifer* (the largest Neotropical dung beetle species), *Oxysternon festivum*, and *Dichotomius boreus*, were sampled at all three sites. These species move long distances and require large, continuous areas of forest to persist. Their presence at the sites is indicative of the intact, contiguous landscape around Kwamala. These large dung beetle species are also the most ecologically important for burying seeds and controlling parasites.

Six species (*Dendropaemon* sp. 1, *Deltorhinum guyanensis*, and four *Anomiopus* species) were only sampled in flight intercept traps and their distinctive morphology, with strongly reduced tarsi and stout, compact bodies, suggest that they are myrmecophilous (associated with ant nests), as are several other dung beetle species. Based on a recent revision of the genus *Anomiopus*, this is the first record for all four of these species in Suriname (Canhedo 2006), although I collected *A. parallelus* and *A. lacordairei* on another RAP survey in Suriname (Larsen 2007; Appendix A). Both species were previously known only from French Guiana and northern Brazil. *Deltorhinum guyanensis*, endemic to the Guianas, was only described after this RAP survey was conducted (Genier 2010), and this is the first record of this species in Suriname.

Deltochilum valgum is a highly specialized predator of millipedes, and adults decapitate and feed on millipedes that are much larger than themselves. This unusual behavior was only discovered and described last year (Larsen et al. 2009). *Canthidium* cf. *chrysis* is a member of the escalerei species group which commonly feed on dead invertebrates. It was captured mostly with dead millipedes, but occasionally with carrion, and may be specialized to feed on millipedes. *Canthidium* sp. 20 (aff. *chrysis*), *Canthon* sp. 1 and *Canthon* sp. 2 were also most abundant at dead millipedes, but whether they specialize on millipedes or on dead invertebrates in general is not yet clear. *Canthidium* cf. *gigas*, which was represented by only one individual in a flight intercept trap, is a member of an unusual species group which may feed on fungus. This group includes by far the largest of all *Canthidium* species. *Canthidium* cf. *minimum* is an unusual species that may need to be transferred to a different genus (see Appendix A for this and other taxonomic notes).

CONSERVATION RECOMMENDATIONS

The Kwamala area supports vast tracts of intact primary forest, which is important for many dung beetle species. Consequently, I found extremely high species richness of dung beetles in the area (94 species). To put this diversity into perspective, during a RAP survey at the Nassau and Lely plateaus in Suriname, I sampled only 24 species and 33 species at each site respectively (Table 3). I sampled extensively in lowland forest around Lago Guri in Bolivar, Venezuela, and found only 41 species (Larsen et al. 2008). On the other hand, small-scale habitat loss and disturbance around Kwamala led to local dung beetle extinctions, which would likely be exacerbated by more widespread habitat loss. Preventing mining operations and other drivers of deforestation from entering the area will be important for maintaining the high biodiversity of the Kwamala region.

In addition to high overall species richness, I found high Beta diversity at the sites across very small spatial scales, and Kutari supports a very distinct dung beetle community than the other sites. Consequently, it is important to protect the diversity of soils and habitats that occur in the Kwamala region even at small spatial scales. Plans for protected areas or reserves should incorporate this small-scale spatial heterogeneity.

Tropical ectotherms, such as dung beetles, are among the most sensitive organisms on Earth to climate change (Larsen et al. 2011). Climate warming is forcing many species to shift their distribution poleward or upslope, and these effects are strongest at the edge of species' ranges. Since the Kwamala area contains many Amazonian species near the edge of their range limit, it may present an excellent opportunity to monitor the response of populations and species' distributions to climate change.

High dung beetle diversity at Kutari, the most isolated site, was correlated with high mammal, including primate, species richness, and this may be explained by lower hunting pressures. The abundance and biomass of dung beetles in the Kwamala area overall was relatively high, and was higher than I observed at Nassau and Lely in other parts of Suriname. This suggests that in addition to the pristine state of the forest, populations of large birds and mammals are relatively stable. However, dung beetle abundance was lower than I expected based on surveys in other Neotropical primary forests where no hunting occurs. This is likely to reflect the relatively low abundance of spider monkeys, howler monkeys, and white-lipped peccaries, which are among the most important species for dung beetles but are also preferred for bushmeat. Reduced hunting on these key species would help to stabilize ecosystem dynamics not just for dung beetles, but for seed dispersal and other ecological processes as well. The establishment of hunting-restricted reserves such as the one at Iwana Samu is an excellent way to maintain sustainable populations of large mammals.

ACKNOWLEDGEMENTS

I would like to thank the Trio People of Kwamalasamutu, the game wardens, and the park guards, as well as Leeanne Alonso and Brian O'Shea for coordinating the RAP survey. Dana Price and Francois Feer provided valuable comparative data from French Guiana.

LITERATURE CITED

Canhedo, V. L. 2006. Revisao taxonomica do genero Anomiopus Westwood, 1842 (Coleoptera, Scarabaeidae, Scarabaeinae). [Taxonomic revision of the genus Anomiopus Westwood, 1842 (Coleoptera, Scarabaeidae, Scarabaeinae).]. Arquivos de Zoologia Sao Paulo 37:349–502.

Colwell, R. K. and J. A. Coddington. 1994. Estimating terrestrial biodiversity through extrapolation. Philosophical Transactions of the Royal Society of London B Biological Sciences 345:101–118.

Edmonds, W. D. and J. Zidek. 2004. Revision of the Neotropical dung beetle genus Oxysternon (Scarabaeidae: Scarabaeinae: Phanaeini). Folia Heyrovskyana Supplementum 11:1–58.

Edmonds, W. D. and J. Zidek. 2010. A taxonomic review of the neotropical genus Coprophanaeus Olsoufieff, 1924 (Coleoptera: Scarabaeidae, Scarabaeinae). Insecta Mundi 0129:1–111.

Feer, F. 2000. Dung and carrion beetles of the rain forest of French Guiana: composition and structure of the guild. Annales De La Societe Entomologique De France 36:29–43.

Gardner, T. A., J. Barlow, I. S. Araujo, T. C. Avila-Pires, A. B. Bonaldo, J. E. Costa, M. C. Esposito, L. V. Ferreira, J. Hawes, M. I. M. Hernandez, M. S. Hoogmoed, R. N. Leite, N. F. Lo-Man-Hung, J. R. Malcolm, M. B. Martins, L. A. M. Mestre, R. Miranda-Santos, W. L. Overal, L. Parry, S. L. Peters, M. A. Ribeiro, M. N. F. da Silva, C. D. S. Motta, and C. A. Peres. 2008a. The cost-effectiveness of biodiversity surveys in tropical forests. Ecology Letters 11:139–150.

Gardner, T. A., M. I. M. Hernandez, J. Barlow, and C. A. Peres. 2008b. Understanding the biodiversity consequences of habitat change: the value of secondary and plantation forests for neotropical dung beetles. Journal of Applied Ecology 45:883–893.

Genier, F. 2009. Le Genre Eurysternus Dalman, 1824 (Scarabaeidae: Scarabaeinae: Oniticellini) Pensoft, Bulgaria.

Genier, F. 2010. A review of the Neotropical dung beetle genera Deltorhinum Harold, 1869, and Lobidion gen. nov. (Coleoptera: Scarabaeidae: Scarabaeinae). Zootaxa 2693:35–48.

Larsen, T. H. 2007. Dung beetles of the Lely and Nassau plateaus, Eastern Suriname. Pages 99–101 *in* L. E. Alonso and J. H. Mol, editors. A rapid biological assessment of the Lely and Nassau plateaus, Suriname (with additional information on the Brownsberg Plateau). Conservation International, Arlington, VA, USA.

Larsen, T. H., F. Escobar, and I. Armbrecht. 2011. Insects of the Tropical Andes: diversity patterns, processes and global change. Pages 228–244 *in* S. K. Herzog, R. Martinez, P. M. Jorgensen, and H. Tiessen, editors. Climate Change and Biodiversity in the Tropical Andes. Inter-American Institute of Global Change Research (IAI) and Scientific Committee on Problems of the Environment (SCOPE), São José dos Campos and Paris.

Larsen, T. H. and A. Forsyth. 2005. Trap Spacing and Transect Design for Dung Beetle Biodiversity Studies. Biotropica 37:322–325.

Larsen, T. H., A. Lopera, and A. Forsyth. 2008. Understanding Trait-Dependent Community Disassembly: Dung Beetles, Density Functions, and Forest Fragmentation. Conservation Biology 22:1288–1298.

Larsen, T. H., A. Lopera, A. Forsyth, and F. Genier. 2009. From coprophagy to predation: a dung beetle that kills millipedes. Biology Letters 5:152–155.

Nichols, E., S. Spector, J. Louzada, T. Larsen, S. Amequita, and M. E. Favila. 2008. Ecological functions and ecosystem services provided by Scarabaeinae dung beetles. Biological Conservation 141:1461–1474.

Scheffler, P. Y. 2005. Dung beetle (Coleoptera : Scarabaeidae) diversity and community structure across three disturbance regimes in eastern Amazonia. Journal of Tropical Ecology 21:9–19.

Spector, S. 2006. Scarabaeine dung beetles (Coleoptera : Scarabaeidae : Scarabaeinae): An invertebrate focal taxon for biodiversity research and conservation. Coleopterists Bulletin 60:71–83.

Appendix A. Dung beetle species abundance (number individuals collected), including taxonomic notes and amended species list from RAP #43.

	Kutari	Sipaliwini	Werehpai	Nassau	Lely	Old species name (from RAP #43 Lely and Nassau)
# Species	70	62	67	27	38	
Total abundance	910	2093	1551	204	906	
# Trap samples	66	61	66	51	53	
Agamopus castaneus Balthasar	0	8	23			
Anomiopus andrei Canhedo	1	0	0			
Anomiopus globosus Canhedo	2	0	0			
Anomiopus lacordairei Waterhouse	3	0	0	1	0	*Anomiopus* sp. 2
Anomiopus parallelus Harold[1]	3	0	1	0	1	*Anomiopus* sp. 1
Ateuchus cereus Harold[2]	2	0	0			
Ateuchus cf. *obscurus* Harold[3]	4	15	9			
Ateuchus cf. *sulcicollis* Harold[4]	1	0	4			
Ateuchus murrayi Harold	27	42	7	1	1	*Ateuchus* sp. 1
Ateuchus pygidialis Harold[5]	1	0	3			
Ateuchus simplex LePeletier & Serville[6]	0	211	74	1	13	*Ateuchus* sp. 2
Ateuchus substriatus Harold	1	12	44			
Ateuchus sp. 3[7]	1	3	1			
Ateuchus sp. 4	0	1	0			
Ateuchus sp. 5[8]	10	7	9			
Ateuchus sp. 6 (aff. *murrayi*)[9]	3	0	0			
Ateuchus sp. 7 (aff. *aeneomicans*)[10]	0	2	2			
Canthidium cf. *chrysis* Fabricius[11]	1	19	2			
Canthidium cf. *gigas* Balthasar[12]	0	0	1			
Canthidium cf. *kirschi* Harold[13]	17	1	1	0	1	*Canthidium* cf. *bicolor*
Canthidium cf. *minimum* Harold[14]	0	2	0			
Canthidium cf. *onitoides* Perty[15]	1	0	0			
Canthidium deyrollei Harold	13	71	32			
Canthidium dohrni Harold[16]	3	4	0			
Canthidium gerstaeckeri Harold	19	12	7	0	6	*Canthidium* sp. 1
Canthidium gracilipes Harold	12	1	3			
Canthidium splendidum Preudhomme de Borre	0	0	11			
Canthidium sp. 5 (aff. *funebre*)[15]	1	4	1			
Canthidium sp. 6[17]	30	3	2	0	4	*Canthidium* sp. 2
Canthidium sp. 7 (aff. *histrio*)	0	2	0			
Canthidium sp. 8 (aff. *quadridens*)	5	4	1			
Canthidium sp. 9	3	2	1			
Canthidium sp. 10	2	0	0			
Canthidium sp. 11 (aff. *guyanense*)[18]	7	0	0			
Canthidium sp. 12 (aff. *latum*)	8	0	5			

table continued on next page

	Kutari	Sipaliwini	Werehpai	Nassau	Lely	Old species name (from RAP #43 Lely and Nassau)
Canthidium sp. 13	3	0	0			
Canthidium sp. 14 (centrale grp)[19]	0	0	2			
Canthidium sp. 15	1	0	1			
Canthidium sp. 16	0	2	0			
Canthidium sp. 17	0	0	1			
Canthidium sp. 18 (aff. *bicolor*)[13]	20	2	3			
Canthidium sp. 19 (aff. *kirschi*)[13]	1	0	0			
Canthidium sp. 20 (aff. *chrysis*)[11]	3	8	7			
Canthon bicolor Castelnau	9	32	31	2	46	*Canthon bicolor*
Canthon quadriguttatus Olivier	1	0	0	1	7	*Canthon quadriguttatus*
Canthon semiopacus Harold	0	1	0			
Canthon sordidus Harold	21	0	12	9	19	*Anisocanthon* cf. *sericinus*
Canthon triangularis Drury	150	155	184	13	14	*Canthon triangularis*
Canthon sp. 1[20]	0	5	1			
Canthon sp. 2[21]	2	4	3			
Canthonella silphoides Harold	0	0	1			
Coprophanaeus jasius Olivier	2	1	2			
Coprophanaeus lancifer Linnaeus	2	1	1	0	1	*Coprophanaeus lancifer*
Coprophanaeus parvulus Olsoufieff	0	1	1	0	1	*Coprophanaeus* cf. *parvulus*
Deltochilum carinatum Westwood	4	0	0	2	2	*Deltochilum carinatum*
Deltochilum guyanense Boucomont	2	3	1	8	0	*Deltochilum* sp. 1
Deltochilum icarus Olivier	3	1	7	1	3	*Deltochilum icarus*
Deltochilum septemstriatum Paulian	4	4	6	4	0	*Deltochilum* sp. 2
Deltochilum valgum Burmeister	3	0	1			
Deltorhinum guyanensis Genier	2	0	0			
Dendropaemon sp. 1	3	0	0			
Dichotomius boreus Olivier	52	123	44	4	7	*Dichotomius* sp. aff. *podalirius*
Dichotomius cf. *lucasi* Harold	38	168	154			
Dichotomius mamillatus Felsche	2	1	1	0	1	*Dichotomius mamillatus*
Dichotomius robustus Luederwaldt	1	1	1			
Dichotomius subaeneus Castelnau	0	1	0			
Dichotomius sp. 2	2	0	1			
Dichotomius sp. 3 (batesi-inachus grp)[22]	0	0	2			
Dichotomius sp. 4	1	2	2			
Dichotomius sp. 5 (calcaratus grp)	1	0	0			
Eurysternus atrosericus Genier	9	35	42			
Eurysternus balachowskyi Halffter & Halffter	0	2	1	1	0	*Eurysternus* sp. 2
Eurysternus cambeforti Genier	0	6	2	0	1	*Eurysternus* cf. *hirtellus*
Eurysternus caribaeus Herbst	21	125	150	5	16	*Eurysternus caribaeus*
Eurysternus cyclops Genier	1	0	0	4	17	*Eurysternus* sp. aff. *caribaeus*

table continued on next page

	Kutari	Sipaliwini	Werehpai	Nassau	Lely	Old species name (from RAP #43 Lely and Nassau)
Eurysternus foedus Guerin-Meneville	4	9	4			
Eurysternus hamaticollis Balthasar	0	2	0			
Eurysternus ventricosus Gill	1	2	0	0	1	*Eurysternus* sp. 1
Hansreia affinis Fabricius	25	53	19	88	569	*Hansreia affinis*
Onthophagus cf. *xanthomerus* Bates[23]	14	6	2			
Onthophagus haematopus Harold	46	589	294	34	52	*Onthophagus* sp. 1
Onthophagus rubrescens Blanchard	134	128	26	1	11	*Onthophagus* cf. *haematopus*
Oxysternon durantoni Arnaud	29	19	8	0	24	*Oxysternon* cf. *durantoni*
Oxysternon festivum Linnaeus	4	9	26			
Oxysternon spiniferum Castelnau	1	1	1			
Phanaeus bispinus Bates	0	1	0			
Phanaeus cambeforti Arnaud	5	5	31			
Phanaeus chalcomelas Perty	40	131	165	2	7	*Phanaeus chalcomelas*
Sulcophanaeus faunus Fabricius	1	0	0			
Sylvicanthon cf. *securus* Schmidt[24]	0	4	1	0	4	*Sylvicanthon* sp. nov.
Trichillum pauliani Balthasar	0	2	21			
Uroxys gorgon Arrow	0	0	1			
Uroxys pygmaeus Harold	23	7	15	4	1	*Uroxys* sp. 2
Uroxys sp. 3	38	15	28	6	29	*Uroxys* sp. 3
Additional species sampled during RAP #43						
Canthidium guyanense Boucomont				2	20	*Canthidium* sp. 4
Canthidium sp. 3				0	3	*Canthidium* sp. 3
Canthon mutabilis Lucas				0	3	*Canthon mutabilis*
Coprophanaeus dardanus MacLeay				0	3	*Coprophanaeus* cf. *dardanus*
Deltochilum orbiculare Lansberge				3	1	*Deltochilum* sp. 3
Dichotomius sp. 1				1	0	*Dichotomius* sp. 1
Eurysternus hypocrita Balthasar				1	1	*Eurysternus velutinus*
Eurysternus vastiorum Martinez				0	2	*Eurysternus* sp. 1
Oxysternon silenus Castelnau				0	2	*Oxysternon aeneum*
Scybalocanthon pygidialis Schmidt				1	10	*Scybalocanthon cyanocephalus*
Uroxys sp. 1				4	2	*Uroxys* sp. 1

[1]Individuals here are larger than *A. parallelus* revised by (Canhedo 2006), and are also larger and differ in pronotal patterning from the individual from the Lely RAP survey

[2]Species needs to be transferred from genus *Canthidium*. *Ateuchus scatimoides* (Balthasar, 1939) is a junior synonym of *Ateuchus cereus*

[3]May match *Canthidium obscurum*, although I have not yet seen this species; if so, species needs to be transferred from genus *Canthidium*

[4]Species needs to be transferred from genus *Canthidium*

[5]Probably represents a species complex; need to study types

[6]The species I collected here matches the type specimen of *Ateuchus setulosus* (Balthasar, 1939); based on museum specimens and the original description, *A. setulosus* appears to be a junior synonym of *A. simplex*, but I have not seen *A. simplex* types.

[7]Similar to *A. pygidialis*, but body more elongate and narrow

[8]Similar to *A. murrayi*, but smaller pygidium with dorsal punctures, among other differences

[9]Similar to *A. murrayi*, but larger, more heavily punctate pygidium, etc. Matches a probably undescribed species I have collected in southeastern Peru at dung and fruit

[10]Smaller than *A. aeneomicans* and with prominent swelling on pygidium which is not present in *A. aeneomicans*

[11]The only obvious difference I can find between *Canthidium* cf. *chrysis* and *Canthidium* sp. 20 (aff. *chrysis*), both of which were collected sympatrically, is color (*C.* sp. 20 is orange and black, while *C.* cf. *chrysis* is green). The aedeagus appears identical. Further study is needed, and these identifications are also based on uncertain museum labels

[12]Part of a species group that needs revision

[13]*Canthidium kirschi* and *Canthidium bicolor* are members of a taxonomically difficult species group which includes several undescribed species. All are very small species with a yellow/orange pronotum, dark brown/black elytra and head, and unarmed head lacking tubercles. Both species, and others, are frequently mixed and misidentified in collections. Key differences include punctures on the pronotum and shape of the male foretibial teeth and claw

[14]This is a curious species. It closely resembles *Canthidium minimum*, although the hind tibia is slightly less curved in the specimens from this survey. *Canthidium minimum* shares characters with two genera, *Canthidium* and *Sinapisoma*, and might need to be transferred to *Sinapisoma*. *Sinapisoma* is currently a monospecific genus, and the only known species possesses a more elongate and curved inner margin of the hind tibia (which until recently caused it to be erroneously considered a canthonine roller) than in *C. minimum*. However, the hind tibia of *C. minimum* is more elongate and curved than other *Canthidium* species. Both share other characters, including a narrow mesosternum.

[15]*Canthidium onitoides* and *Canthidium funebre* are members of a species complex whose species are frequently misidentified in collections and needs further revision. I have seen the *C. funebre* type, which is from Suriname, and it has microsculptured, matte elytra and yellow femora, in contrast to other species with shining, glabrous elytra and/or unicolor legs. *Canthidium* cf. *onitoides* collected during this RAP has glabrous elytra, and needs to be compared with *C. onitoides* type.

[16]A similar species from southeastern Peru has two long fovea along the posterior elytral striae, rather than three as in the species collected here. It's unclear which of these two species is actually *Canthidium dohrni*; the type is from Para, Brazil

[17]Matches a possibly undescribed species collected in Colombia

[18]Very similar to *Canthidium guyanense*, which was collected during Nassau and Lely RAP surveys, but can be separated based on the second and third elytral striae which are not deeply impressed posteriorly

[19]Matches a possibly undescribed species I have collected in SE Peru. Perhaps the smallest member of the lentum-centrale species group

[20]Very similar to *Canthon* sp. 2 (possibly same species), but pronotum appears more glabrous and shining, with less microsculpturing

[21]Matches a species currently being described, *Canthon doesburgi* (Huijbregts, in litt.), but no name is yet available and the species remains formally undescribed

[22]Matches a possibly undescribed species from SE Peru. Possesses an unusual fovea on the posterior portion of the head

[23]*O. xanthomerus* is part of a difficult species group (clypeatus species group), that needs revision. *O. xanthomerus* usually has dark legs with yellow femora, although the species collected here has dark, unicolor legs. *O. clypeatus* has dark legs, but the male pronotal carinae are much sharper and more pronounced, while they are relatively smooth and rounded in *O. xanthomerus*.

[24]I have not seen any specimens of *Sylvicanthon securus*, but the species collected here appears to match the original description, and the type locality is Suriname. I have often seen this species misidentified as *Sylvicanthon candezei*, but *S. candezei* has 2 foretibial teeth rather than 3 as in the species here

Appendix B. Diet preference/capture method for dung beetles. Data are number of individuals collected.

	Dung	Carrion	Dead millipedes	Injured millipedes	Fungus	FIT
# Species	67	21	4	2	2	58
Total abundance	4123	105	26	2	7	290
# Trap samples	124	17	4	2	8	38
Agamopus castaneus Balthasar	31					
Anomiopus andrei Canhedo						1
Anomiopus globosus Canhedo						2
Anomiopus lacordairei Waterhouse						3
Anomiopus parallelus Harold						4
Ateuchus cereus Harold	1	1				
Ateuchus cf. *obscurus* Harold	28					
Ateuchus cf. *sulcicollis* Harold	2					3
Ateuchus murrayi Harold	69					7
Ateuchus pygidialis Harold	3					1
Ateuchus simplex LePeletier & Serville	282	1				2
Ateuchus substriatus Harold	52	2				3
Ateuchus sp. 3	5					
Ateuchus sp. 4						1
Ateuchus sp. 5	7	3			4	12
Ateuchus sp. 6 (aff. *murrayi*)						3
Ateuchus sp. 7 (aff. *aeneomicans*)						4
Canthidium cf. *chrysis* Fabricius		3	15			4
Canthidium cf. *gigas* Balthasar						1
Canthidium cf. *kirschi* Harold	1					18
Canthidium cf. *minimum* Harold						2
Canthidium cf. *onitoides* Perty						1
Canthidium deyrollei Harold	114	2				
Canthidium dohrni Harold	5					2
Canthidium gerstaeckeri Harold	37					1
Canthidium gracilipes Harold	1					15
Canthidium splendidum Preudhomme de Borre	11					
Canthidium sp. 5 (aff. *funebre*)	5					1
Canthidium sp. 6	34					1
Canthidium sp. 7 (aff. *histrio*)						2
Canthidium sp. 8 (aff. *quadridens*)						10
Canthidium sp. 9	5					1
Canthidium sp. 10	2					
Canthidium sp. 11 (aff. *guyanense*)	5					2
Canthidium sp. 12 (aff. *latum*)						13
Canthidium sp. 13						3
Canthidium sp. 14 (centrale grp)	1					1
Canthidium sp. 15						2

table continued on next page

	Dung	Carrion	Dead millipedes	Injured millipedes	Fungus	FIT
Canthidium sp. 16	2					
Canthidium sp. 17	1					
Canthidium sp. 18 (aff. *bicolor*)	5					20
Canthidium sp. 19 (aff. *kirschi*)						1
Canthidium sp. 20 (aff. *chrysis*)		7	4	1		6
Canthon bicolor Castelnau	68					4
Canthon quadriguttatus Olivier	1					
Canthon semiopacus Harold	1					
Canthon sordidus Harold	20	13				
Canthon triangularis Drury	469	18				2
Canthon sp. 1		1	3			2
Canthon sp. 2		2	4			3
Canthonella silphoides Harold						1
Coprophanaeus jasius Olivier	2	3				
Coprophanaeus lancifer Linnaeus	4					
Coprophanaeus parvulus Olsoufieff						2
Deltochilum carinatum Westwood	4					
Deltochilum guyanense Boucomont	3	3				
Deltochilum icarus Olivier	9	2				
Deltochilum septemstriatum Paulian	1	13				
Deltochilum valgum Burmeister						4
Deltorhinum guyanensis Genier						2
Dendropaemon sp. 1						3
Dichotomius boreus Olivier	219					
Dichotomius cf. *lucasi* Harold	305	12		1	3	39
Dichotomius mamillatus Felsche	4					
Dichotomius robustus Luederwaldt	3					
Dichotomius subaeneus Castelnau	1					
Dichotomius sp. 2	1					2
Dichotomius sp. 3 (batesi-inachus grp)	2					
Dichotomius sp. 4	4					1
Dichotomius sp. 5 (calcaratus grp)	1					
Eurysternus atrosericus Genier	77	8				1
Eurysternus balachowskyi Halffter & Halffter	3					
Eurysternus cambeforti Genier	7	1				
Eurysternus caribaeus Herbst	293	3				
Eurysternus cyclops Genier	1					
Eurysternus foedus Guerin-Meneville	17					
Eurysternus hamaticollis Balthasar	2					
Eurysternus ventricosus Gill	3					
Hansreia affinis Fabricius	97					
Onthophagus cf. *xanthomerus* Bates	15	5				2

table continued on next page

	Dung	Carrion	Dead millipedes	Injured millipedes	Fungus	FIT
Onthophagus haematopus Harold	920					9
Onthophagus rubrescens Blanchard	285					3
Oxysternon durantoni Arnaud	56					
Oxysternon festivum Linnaeus	36					3
Oxysternon spiniferum Castelnau						3
Phanaeus bispinus Bates	1					
Phanaeus cambeforti Arnaud	36					5
Phanaeus chalcomelas Perty	332					4
Sulcophanaeus faunus Fabricius	1					
Sylvicanthon cf. *securus* Schmidt	4					1
Trichillum pauliani Balthasar	22					1
Uroxys gorgon Arrow	1					
Uroxys pygmaeus Harold	37					8
Uroxys sp. 3	47	2				32

Chapter 6

A rapid biological assessment of katydids of the Kwamalasamutu region, Suriname (Insecta: Orthoptera: Tettigoniidae)

Piotr Naskrecki

SUMMARY

Seventy-eight species of katydids (Orthoptera: Tettigoniidae) were recorded during a rapid biological assessment of lowland forests of the Kwamalasamutu Region, Suriname. At least seven species are new to science, and 29 species are recorded for the first time from Suriname, bringing the number of species of katydids known from this country up to 85. The current survey confirms that the katydid fauna of Suriname is exceptionally rich, yet still very poorly known. Although no specific conservation issues have been determined to affect the katydid fauna, habitat loss in Suriname due to logging and mining activities constitute the primary threat to the biota of this country.

INTRODUCTION

Katydids (Insecta: Orthoptera: Tettigoniidae) have long been recognized as organisms with a significant potential for use in conservation practices. Many katydid species exhibit strong microhabitat fidelity, low dispersal abilities (Rentz 1993), and high sensitivity to habitat fragmentation (Kindvall and Ahlen 1992) thereby making them good indicators of habitat disturbance. These insects also play a major role in many terrestrial ecosystems as herbivores and predators (Rentz 1996). It has been demonstrated that katydids are a principal prey item for several groups of invertebrates and vertebrates in Neotropical forests, including birds, bats (Belwood 1990), and primates (Nickle and Heymann 1996). While no Neotropical katydids have been classified as threatened (primarily because of the paucity of data on virtually all species known from this region), there are already documented cases of some Nearctic katydids becoming threatened or endangered, or even extinct (Rentz 1977.)

Despite the recent increase in the faunistic and taxonomic work on katydids of the Neotropics, forests of the Guiana Shield remain some of the least explored and potentially interesting areas of South America. Collectively, over 190 species of the Tettigoniidae have been recorded from countries comprising the Guiana Shield (e.g., Venezuela, Guyana, Suriname, and French Guiana), but this number clearly represents a small fraction of the regional species diversity, and at least 300–500 species can be expected to occur there. Fifty-six species have been reported from Suriname (Eades et al. 2011). Virtually all of these records are based on material collected in the 19th century, and no targeted survey of the katydid fauna of the country has ever been conducted. Most of the species from Suriname were described in the monographic works by Brunner von Wattenwyl (1878, 1895), Redtenbacher (1891), and Beier (1960, 1962). More recently Nickle (1984), Emsley and Nickle (2001), Kevan (1989), and Naskrecki (1997) described additional species from the region.

The following report presents preliminary results of a survey of katydids conducted between 17 August and 9 September 2010 at selected sites in the Kwamalasamutu region of southern Suriname.

METHODS AND STUDY SITES

During the survey, three methods were employed for collecting katydids: (1) collecting at an ultraviolet (UV) light at night, (2) visual searching at night and during the day, and (3) detection of stridulating individuals using an ultrasound detector (Petersson 200) at night. Representatives of all encountered species were collected and voucher specimens were preserved in 95% ethanol or as dry specimens layered between thin paper tissue and desiccated with silica gel. Voucher specimens of all collected species will be deposited in the National Zoological Collection of Suriname, while remaining specimens will be deposited in the collections of the Museum of Comparative Zoology at Harvard University, and the Academy of Natural Sciences of Philadelphia (the latter will also become the official repository of the types of new species encountered during the present survey upon their formal description.)

In addition to physical collection of specimens, stridulation of acoustic species was recorded using a Marantz PMD661 digital recorder with a Sennheiser directional microphone. Virtually all species encountered were photographed, and these images will be available online in the database of the world's katydids (Eades et al. 2011).

Simpson's Index of Diversity (Ds) was calculated for each site using the formula:

$$D_s = 1 - \sum_1^i [n_i(n_i-1)]/[N(N-1)]$$

where ni = number of individuals of species i, and N = number of all collected individuals.

Katydids were surveyed at the following five sites:

1. **Camp 1: Kutari, Site 1** (2°10'31.3"N, 56°47'14.1"W); 18–25 August 2010

2. **Iwana Samu** (2°21'46.6"N, 56°45'17.9"W); 25–26 August 2010

3. **Camp 2: Sipaliwini, Site 2** (2°17'24.1"N, 56°36'25.6"W); 27 August – 2 September 2010

4. **Inselberg nr. Sipaliwini river** (2°17'56.4"N, 56°36'37.3"W); 31 August 2010

5. **Camp 3: Werehpai, Site 3** (2°21'47.1"N, 56°41'51.5"W); 2–8 September 2010

RESULTS

The katydid fauna documented during this survey was exceptionally interesting and rich in species. Seventy-eight species were recorded during the survey, of which at least seven species are new to science, and one of these will likely be placed in a new genus of sylvan katydids. Twenty-nine species are recorded for the first time from Suriname, bringing the total number of species known from this country to 85 (a 52% increase). A complete checklist of species collected during the survey is presented in the Appendix.

It is worth mentioning that the abundance of katydids encountered during this survey was often exceptionally low (although no formal structured sampling was conducted, the rate of katydid collection was often only one individual/hour, and during most nights no individuals were attracted to the UV light.) This low abundance is well reflected in the species richness indices of the three main camps, with Simpson's Indexes of Diversity (Ds) for Camps 1, 2, and 3 being 0.969030969, 0.999625047, and 0.997416324, respectively. These results indicate very high species richness, combined with low abundance of individual species, a situation typical of tropical habitats with low levels of disturbance.

Many of the recorded species appeared only as nymphs, often in early developmental stages, which indicates a strong seasonality in their development. It seems that in such species egg hatching must take place in the last weeks of the rainy season, and maturation takes place during the dry season.

Of the three main camps, the first site (Kutari) had the lowest number of both species (25) and specimens (78) collected, presumably because of the heavy rains that still affected the activity of katydids at the end of the rainy season, when the survey began. Werehpai had the highest number of species (54), followed by Sipaliwini (46).

Below I discuss the most interesting taxa of katydids recorded during the survey.

Conehead katydids (subfamily Conocephalinae)

The Conocephalinae, or the conehead katydids, include a wide range of species found in both open, grassy habitats, and high in the forest canopy. Many species are obligate semenivores (seed feeders), while others are strictly predaceous. A number of species are diurnal, or exhibit both diurnal and nocturnal patterns of activity. Sixteen species of this family were recorded.

Vestria sp. n. — Four species of this genus are known from lowland forests of Central and South America. These insects, known as Crayola katydids because of their striking coloration, are the only katydids known to employ chemical defenses, which are effective at repelling bird and mammalian predators (Nickle et al. 1996). Specimens of *Vestria* collected at Sipaliwini and Werehpai represent a species new to science.

Eschatoceras sp. 1 — A single female specimen of this species was retrieved from a spider web at Werehpai. The insect was already partially digested, but its diagnostic characters, such as the unique shape of the fastigium of vertex and the subgenital plate, allow me to conclude that it most likely represents a yet unnamed species of this genus.

Subria cf. *amazonica* — Ten specimens of this species were recorded at the Sipaliwini and Werehpai camps. They resemble *S. amazonica* Redtenbacher, a species known from the unique female holotype from "Alto Amazonas", but differs in the degree of the development of the wings, and likely represent a yet unnamed species of the genus. These predaceous insects were represented by two distinct, green and orange, color morphs.

Loboscelis baccatus Nickle & Naskrecki — This arboreal, most-likely predaceous species was previously known only from Amazonian Peru (Nickle and Naskrecki 2000), but a single individual was found at Werephai. This record represents a significant extension of its range, and the first record of the genus *Loboscelis* in the Guiana Shield.

Leaf katydids (subfamily Phaneropterinae)

The Phaneropterinae, or leaf katydids, represent the largest, most species-rich lineage of katydids, with nearly 2,700 species worldwide, and at least 550 species recorded from South America. All species of this family are obligate herbivores, often restricted to a narrow range of host plants. Probably at least 50–75% of species found in lowland rainforests are restricted to the canopy layer and never descend to the ground (females of many species lay eggs on the surface of leaves or stems, and the entire nymphal development takes place on a single host plant.) For this reason, these insects are difficult to collect, and the only reliable method for their collection is a UV or mercury-vapor lamp, or canopy fogging. Very few species can be encountered during a visual or acoustic search in the understory of the forest.

Twenty-five species of leaf katydids were recorded during the present survey, virtually all attracted to the UV light at the camps.

Euceraia sp. n. — A single individual of an undescribed species of this genus was collected at the Sipaliwini camp at the UV light at night. It is a member of a genus of canopy katydids, known to deposit their eggs between the layers of the leaf epidermis; they are rarely encountered at the understory level of the forest.

Meneghelia carlotae Piza — This is the first record of this species from Suriname and the first specimen collected since its original description (Piza 1980) from "Territ. do Amapá" in Brazil. The morphology of the female ovipositor in this species is unique among katydids, and suggests that the eggs are laid in some unusual, yet unknown substrate.

Polichnodes americana Giglio-Tos — Two individuals of this species were collected at the Sipaliwini camp at the UV light. They represent the first records of this species outside its type locality in Ecuador and a substantial extension of its range; they are also the first specimens collected since its original description over a century ago (Giglio-Tos 1898).

Sylvan katydids (subfamily Pseudophyllinae)

Virtually all members of tropical Pseudophyllinae occur only in forested, undisturbed habitats, and thus have a potential as indicators of habitat changes. These katydids are mostly herbivorous, although opportunistic carnivory has been observed in some species (e. g., *Panoploscelis*). Many are confined to the upper layers of the forest canopy and never come to lights, and are therefore difficult to collect. Fortunately, many species have very loud, distinctive calls, and it is possible to document their presence based on their calls alone, a technique well known to ornithologists. Thirty-five species of this family were collected during the present survey.

Gnathoclita vorax (Stoll, 1813) — This spectacular species is a rare example of a katydid with strong sexual dimorphism manifested in strong, allometric growth of the male mandibles. It was found at Werephai, although all collected specimens were nymphal. This species is known only from southern Guyana and southern Suriname.

Eubliastes cf. *adustus* — Three individuals of this large katydid species were collected at Sipaliwini. Although superficially similar to *E. adustus* Bolivar known from Ecuador, the morphology of the male external genitalic structures indicates that these specimens may represent a species new to science. If these specimens do represent *E. adustus*, this would be the first record of this species anywhere outside of Ecuador.

Gen._Homalaspidini sp. 1 — This highly unusual katydid was collected from among leaves of yucca plants growing on the inselberg near the Sipaliwini camp, and another individual was found on similar vegetation at the Werehpai camp. The body of this insect is extremely elongated, adapted to live among long, stiff leaves of yucca and related plants.

CONSERVATION RECOMMENDATIONS

The results of this survey confirm that the fauna of katydids of southern Suriname is exceptionally rich, even by the standards of lowland tropical forests, and that a large proportion of it remains unknown and unnamed. More sampling surveys, combined with comprehensive taxonomic and phylogenetic reviews, are badly needed in order to understand its true magnitude.

As with most groups of tropical insects, the principal threat to the survival of katydids in Suriname comes from habitat loss, especially from logging and mining. While species-level conservation recommendations are currently impossible to make, protecting the existing habitats, or at least major, connected fragments of them, is the most effective way of ensuring their survival.

REFERENCES

Beier, M. 1960. Orthoptera Tettigoniidae (Pseudophyllinae II). – In: Mertens, R., Hennig, W. & Wermuth, H. [eds]. Das Tierreich. – 74: 396 pp.; Berlin (Walter de Gruyter & Co.).

Beier, M. 1962. Orthoptera Tettigoniidae (Pseudophyllinae I). – In: Mertens, R., Hennig, W. & Wermuth, H.

[eds]. Das Tierreich. – 73: 468 pp.; Berlin (Walter de Gruyter & Co.).

Belwood, J.J. 1990. Anti-predator defences and ecology of neotropical forest katydids, especially the Pseudophyllinae. – Pages 8–26 in: Bailey, W.J. & Rentz, D.C.F. [eds]. The Tettigoniidae: biology, systematics and evolution: ix + 395 pp.; Bathurst (Crawford House Press) & Berlin et al. (Springer).

Brunner von Wattenwyl, C. 1878. Monographie der Phaneropteriden. 1–401, pls 1–8; Wien (Brockhaus).

Brunner von Wattenwyl, C. 1895. Monographie der Pseudophylliden. IV + 282 pp. [+ X pls issued separately]; Wien (K.K. Zoologisch–Botanische Gesellschaft).

Eades, D.C.; D. Otte; M.M. Cigliano & H. Braun. Orthoptera Species File Online. Version 2.0/4.0. [25 June 2011]. <http://Orthoptera.SpeciesFile.org>

Emsley, M.G. & Nickle, D.A. 2001. New species of the Neotropical genus *Daedalellus* Uvarov (Orthoptera: Tettigoniidae: Copiphorinae). – Transactions of the American Entomological Society 127: 173–187.

Giglio-Tos, E. 1898. Viaggio del Dr. Enrico Festa nella Republica dell'Ecuador e regioni vicine. Bollettino del Musei di Zoologia ed Anatomia Comparata della R. Universita di Torino, 13 (311): 1–108.

Kevan, D.K.McE. 1989. A new genus and new species of Cocconotini (Grylloptera: Tettigonioidea: Pseudophyllidae: Cyrtophyllinae) from Venezuela and Trinidad, with other records for the tribe. – Bol. Ent. Venez. 5: 1–17.

Kindvall, O. & Ahlen, I. 1992. Geometrical factors and metapopulation dynamics of the bush cricket, *Metrioptera bicolor* Philippi (Orthoptera: Tettigoniidae). – Conservation Biology 6: 520–529.

Naskrecki, P. 1997. A revision of the neotropical genus *Acantheremus* Karny, 1907 (Orthoptera: Tettigoniidae: Copiphorinae). – Transactions of the American Entomological Society 123: 137–161.

Nickle, D.A. 1984. Revision of the bush katydid genus *Montezumina* (Orthoptera; Tettigoniidae; Phaneropterinae). – Transactions of the American Entomological Society 110: 553–622.

Nickle, D.A., J.L. Castner, S.R. Smedley, A.B. Attygalle, J. Meinwald and T. Eisner. 1996. Glandular Pyrazine Emission by a Tropical Katydid: An Example of Chemical Aposematism? (Orthoptera: Tettigoniidae: Copiphorinae: *Vestria* Stål). Journal of Orthoptera Research, 5: 221–223.

Nickle, D.A. & Heymann E.W. 1996. Predation on Orthoptera and related orders of insects by tamarin monkeys, *Saguinus mystax* and *S. fuscicollis* (Primates: Callitrichidae), in northeastern Peru. – Journal of the Zoological Society 239: 799–819.

Nickle D.A. and P. Naskrecki. 2000. The South American Genus *Loboscelis* Redtenbacher, 1891 (Orthoptera: Tettigoniidae: Copiphorinae sensu lato). Jour. Orth. Res., 8: 147–152.

Piza Jr., S. De Toledo. 1980. Oito novos gêneros de Phaneropterinae do Brasil (Orthoptera - Tettigoniidae). Revista de Agricultura (Piracicaba), 55(4): 221–230.

Redtenbacher. 1891. Monographie der Conocephaliden. Verh. der Zoologisch-botanischen Gesellsch Wien 41(2): 315–562.

Rentz, D.C.F. 1977. A new and apparently extinct katydid from antioch sand dunes (Orthoptera: Tettigoniidae). – Entomological News 88: 241–245.

Rentz, D.C.F. 1993. Orthopteroid insects in threatened habitats in Australia. – Pages 125–138 in: Gaston, K.J., New, T.R. & Samways, M.J. [eds]. Perspectives on Insect conservation: 125–138; Andover, Hampshire (Intercept Ltd).

Rentz, D.C.F. 1996. Grasshopper country. The abundant orthopteroid insects of Australia. Orthoptera; grasshoppers, katydids, crickets. Blattodea; cockroaches. Mantodea; mantids. Phasmatodea; stick insects: i–xii, 1–284; Sydney (University of New South Wales Press).

Appendix. List of katydids (Orthoptera: Tettigoniidae) recorded during the Kwamalasamutu RAP survey.

Species	Camp 1 (Kutari)	Camp 2 (Sipaliwini)	Camp 3 (Werehpai)	Iwana Samu	Inselberg nr. Camp 2	New for Suriname	New to science
Conocephalinae							
Agraecia viridipennis		x	x			x	
Eschatoceras bipunctatus			x				
Eschatoceras sp. 1			x				x
Subria cf. *amazonica*		x	x				x
Subria grandis		x	x			x	
Uchuca amacayaca			x			x	
Uchuca sp. 1	x	x	x	x			x
Conocephalus (X.) *cinereus*					x	x	
Acantheremus elegans		x					
Acantheremus sp. 1		x	x				
Copiphora longicauda	x	x	x				
Graminofolium castneri	x	x	x	x		x	
Loboscelis bacatus			x			x	
Neoconocephalus punctipes			x				
Neoconocephalus sp. 2			x		x		
Vestria sp. 1	x	x	x				
Listroscelidinae							
Phlugis teres			x				
Listroscelis sp. 1	x	x	x				
Phaneropterinae							
Dysonia (D.) *fuscifrons*	x					x	
Steirodon (P.) *dentatum*		x	x			x	
Anaulacomera sp. 1	x	x	x				
Anaulacomera sp. 2	x	x	x				
Anaulacomera sp. 5			x				
Anaulacomera sp. 6			x				
Anaulacomera sp. 7			x				
Anaulacomera spatulata			x			x	
Euceraia atryx			x				
Euceraia sp. 1		x					x
Hetaira smaragdina		x	x				
Hyperphrona gracilis		x	x			x	
Ligocatinus cf. *punctatus*			x				x
Meneghelia carlotae			x			x	
Microcentrum marginatum		x				x	
Microcentrum sp. 1		x					
Microcentrum sp. 2		x	x				
Parableta sp. 1		x					
Phylloptera festae		x				x	
Phylloptera laevis	x	x					
Phylloptera sp. 1	x	x					

table continued on next page

Species	Camp 1 (Kutari)	Camp 2 (Sipaliwini)	Camp 3 (Werehpai)	Iwana Samu	Inselberg nr. Camp 2	New for Suriname	New to science
Phylloptera sp. 3		x	x				
Polichnodes americana		x				x	
Theia unicolor		x				x	
Viadana sp. 1	x	x	x				
Pseudophyllinae							
Bliastes contortipes			x				
Eubliastes adustus		x				x	
Meroncidius sp. 1	x						
Schedocentrus (S.) *basalis*		x					
Schedocentrus (S.) *vicinus*			x				
Gnathoclita vorax			x				
Panoploscelis scudderi	x	x	x			x	
Gen_Homalaspidini sp. 1			x		x		x
Chondrosternum sp. 1	x	x	x	x			
Chondrosternum triste	x	x	x	x		x	
Leptotettix falconarius		x	x				
Platychiton surinamus	x	x					
Aemasia sp. 1	x	x					
Platyphyllum sp. 2	x						
Triencentrus amoenus		x				x	
Triencentrus nigrospinosus			x			x	
Triencentrus sp. 1		x					x
Acanthodis sp. 1	x	x	x				
Acanthodis unispinulosa		x	x				
Ancistrocercus truncatistylus			x			x	
Diacanthodis granosa	x					x	
Leurophyllum consanguineum			x			x	
Rhinischia regimbarti			x			x	
Sphyrophyllum malleolatum		x				x	
Cycloptera speculata	x						
Pterochroza ocellata	x	x	x				
Roxelana crassicornis			x			x	
Typophyllum erosum			x			x	
Typophyllum sp. 1	x	x	x				
Typophyllum sp. 2		x	x				
Diophanes salvifolius		x	x			x	
Eumecopterus incisus	x	x	x				
Teleutias aduncus	x					x	
Teleutias surinamus		x					
Teleutias vicinissimus			x	x		x	
Total	**25**	**46**	**54**	**5**	**3**	**29**	**7**

Chapter 7

A preliminary survey of the ants of the Kwamalasamutu region, SW Suriname

Leeanne E. Alonso

SUMMARY

Over 100 species of ants (Hymenoptera: Formicidae) were recorded around the Werehpai caves during the RAP biological assessment of the Kwamalasamutu Region, Suriname in September 2010. While analysis of the ant data is ongoing, preliminary results indicate that the forests around Kwamalasamutu contain a diverse and abundant ant fauna. The presence of many dacetine species typical of closed-canopy rainforest indicates that the forests are in good condition. The ant fauna of Suriname is still very poorly known, as few locations have been sampled for ants. Data on the ant fauna of the Kwamalasamutu area are valuable for eco-tourism and can help to inform tourists about the hidden fauna of the rainforest and their important roles in ecosystem function and conservation.

INTRODUCTION

When people think of the biodiversity of a tropical rainforest, many think first of the colorful parrots and macaws, the elusive yet alluring jaguars and ocelots, and the majestic towering tropical trees. However, the majority of biodiversity in a tropical forest lies in the hidden and overlooked fauna of invertebrates. Ants in particular make up over 15% of the biomass of animals in a tropical forest (Fittkau and Klinge 1973) due to their high abundance. With over 12,000 described species of ants in the world, and their social lifestyle consisting of colonies ranging in size from just a few individuals to millions of workers, ants are a dominant force in all terrestrial ecosystems, especially tropical rainforests. Due in part to their social nature, ants play many critical roles in the functioning of the tropical terrestrial ecosystem, including dispersing seeds, tending mutualistic Homoptera, defending plants, preying on other invertebrates and small vertebrates, and modifying the soil by adding nutrients and aeration (Philpott et al. 2010). Another critical function provided by ants is that of scavenging; ants are often the first animals to arrive upon a dead animal and start the decomposition process. Ants are particularly important to plants since they move soil along the soil profile through the formation of their mounds and tunnels, which directly and indirectly affects the energy flow, habitats, and resources for other organisms (Folgarait 1998).

In addition to their ecological importance, ants have several features that make them especially useful for conservation planning, including: 1) they are dominant members of most terrestrial environments; 2) they are easily sampled in sufficiently high numbers for statistical analysis in short periods of time (Agosti et al. 2000); 3) they are sensitive to environmental change (Kaspari and Majer 2000); and 4) they are indicators of ecosystem health and of the presence of other organisms, due to their numerous symbioses with plants and animals (Alonso 2000).

Ants are also useful organisms for the promotion of eco-tourism. Much of eco-tourism focuses on sightings of birds and large mammals, which are elusive and often very hard to find or see in the dense rainforest. Ants, on the other hand, are ubiquitous and can be seen as soon as one sets foot in almost any forest around the world. Ant behavior and ecology are fascinating, and local guides who can share ant stories will attract and educate many eco-tourists.

STUDY SITE

Ants were studied only at the third RAP site, Werehpai (camp at N 02° 21' 47", W 056° 41' 52"), from 4–7 September 2010. The camp was located on the north bank of the Sipaliwini River on an abandoned farm, and the habitat immediately surrounding the camp was mostly tall second-growth forest with a dense understory. Further from camp along the 3.5-km trail to the Werehpai caves, the habitat was primarily tall terra firme forest, with a few palm swamps and many spiny palms (*Astrocaryum sciophilum*) in the understory. Ants were sampled in the forests along the main trail between camp and the Werehpai caves and along short side trails that ran off the main trail.

METHODS

Ants from the ground, leaf litter, and vegetation were sampled by hand collecting along the main trail and into the forest, as well as around camp. Ants from the leaf litter were also sampled using two sifting methods. The first method was the "David sifter", a plastic tray (20 cm × 30 cm) with an imbedded mesh screen placed over another plastic tray. Leaf litter was collected by hand (using gloves) and placed on the mesh screen. The tray was then shaken back and forth to move ants and other small invertebrates out of the leaf litter, which then fell through the screen and into the tray below. Ants were then collected out of the bottom tray using forceps.

The second sifting method used was the Ants of the Leaf Litter (ALL) protocol (Agosti et al. 2000). Four one-hundred-meter linear transects were sampled at the following locations: 1) near the Werehpai caves (~10:00h, warm and sunny, fairly dry), 2) at the botanical team's 1-ha plot (see plant chapter for coordinates; ~14:00h, warm and sunny, dry), 3) along the trail between camp and the caves (N 02 22'06" W 56 41'23.7", 09:00h), and 4) perpendicular to the main trail (N 02 22'04.9" W56 41'25.1", 11:00h). Along each transect, a 1x1-m quadrat was set up every 10 m (for a total of 10 quadrats per transect). The leaf-litter, rotten twigs, and first layer of soil present in the quadrat were collected into a cloth sifter and shaken for about a minute. Within the sifter was a wire sieve of 1-cm² mesh size which allowed small debris and invertebrates such as ants to fall through the mesh into the bottom of the sifting sack. The sifted leaf litter was then placed in a full-sized Winkler sack, which is a cotton bag into which four small mesh bags containing the leaf litter are placed. Due to their high level of activity, ants run out of the litter and the mesh bag and fall to the bottom of the sack into a collecting cup of 95% ethanol. The Winkler sacks were hung in the field lab for 48 hours. The ant specimens were then preserved in 95% ethanol and sorted to morphospecies. Specimens were identified to species level, when possible, by L. Alonso and J. Sosa-Calvo using ant taxonomic literature, AntWeb (www. antweb.org), and the ant collection at the National Museum of Natural History in Washington, D.C.

RESULTS

Due to the high diversity and large sample size of ants collected, the ant samples were still being processed at the time of this publication. However, preliminary results based on many of the hand collecting samples and the ALL transect sampled near the Werehpai caves indicate a high diversity of ant species typical of a pristine lowland tropical forest, with 105 ant species documented so far. The ALL transect revealed at least 62 ant species, including many species of the tribe Dacetini, and many species of the tribe Attini (fungus-growing ants; Appendix). The most species-rich genus within the ALL transect was *Pheidole*, with at least 20 species, which is consistent with most tropical studies. Other species-rich genera included *Solenopsis*, *Pachycondyla*, and *Odontomachus*. The ants from three of the ALL transects have yet to be sorted and identified; thus more species are likely to be added to the list when these samples have been processed.

Preliminary analysis of some of the hand-collected Davis sifter samples revealed an additional 44 species of ants, many of which were not collected in the litter sample due to their arboreal habits or their ability to escape rapidly when pursued. These included many large ants that were commonly seen in the forest, including the arboreal species *Daceton armigerum*, *Cephalotes* spp., and *Camponotus* spp., the large-eyed terrestrial *Gigantiops destructor*, and several species of army ants (Appendix). Many ant-plants were found in the area, including *Triplaris* sp. and many *Cecropia* sp., which house obligate ant mutualists. *Pseudomyrmex* sp. was collected from *Triplaris* near the RAP camp, and *Azteca* sp. was collected from *Cecropia* along the Sipaliwini River.

Some of the ant species collected are likely new to science and/or new records for Suriname. However, due to the ongoing process of identifying the specimens from the RAP survey, this information is not yet available, but will be published at a later date.

DISCUSSION

Few previous studies of the ants of Suriname have been conducted. Thirty-six ant species were reported by Borgmeier (1934) from coffee plantations around Paramaribo. Kempf (1961) recorded 171 species (54 genera) from primary forest, plantations, and pastures. Most of the ant collections in the interior of Suriname were made by G. Geyskes sporadically between 1938–1958, in Paramaribo and Brownsberg Nature Park.

Prior to the RAP survey of the Lely and Nassau plateaus in eastern Suriname in 2005 (Sosa-Calvo 2007), a total of 290 ant species had been recorded for Suriname (Kempf 1972, Fernandez and Sendoya 2004). Sosa-Calvo (2007) documented a total of 169 species from Lely and Nassau Plateaus, at least half of which are probably new records for Suriname. Thus, 370 ant species is a conservative estimate of the number recorded in Suriname so far. However, given the low effort of ant sampling in Suriname and the few localities sampled, there are likely many more ant species present in Suriname. Tropical lowland rainforests typically harbor a high diversity of ants. For example, Longino et al. (2002) found over 450 ant species in an area of approximately 1500 ha in Costa Rica, and LaPolla et al. (2007) reported 230 species from eight sites in Guyana. A recent RAP survey in Papua New Guinea (Lucky et al. 2011) reported 177 ant species from the lowland site (500 m). Long-term surveys of several tropical faunas regularly record new ant species over time, demonstrating that ants are typically undersampled (i.e. Brühl et al. 1998, Fisher 2005). More studies of ant diversity throughout Suriname are needed to estimate the country's ant diversity, and thereby provide important baseline data for conservation and management of Suriname's biodiversity.

The genus *Acanthomyops* was common in the leaf litter samples but was only recently recorded in Suriname (Sosa-Calvo 2007). Based on its distribution, this genus would be expected to be present in Suriname. Its documentation on the RAP survey highlights the need for continued sampling of ants within Suriname.

With a high diversity and visibility at the Werehpai RAP site, ants certainly play many important roles in the ecosystem. *Pseudomyrmex* and *Azteca* ants protect their host trees (*Triplaris* sp. and *Cecropia* sp., respectively) from herbivores. High ant activity in the leaf litter suggests that ants are playing a key role in scavenging, soil turnover, and predation. Ants also serve as a key food source for many other rainforest animals, including frogs, snakes, small mammals, birds, and other invertebrates. Ants have been found to be the source of the poison in many poison dart frogs. Army ants in particular play a key role as predators in tropical lowland rainforest. Army ant species, such as *Eciton burchelli*, conduct large swarm raids in which millions of workers spread out through the forest, capturing everything they can to bring back to their colony as food. These ants are key to keeping populations of many invertebrates in check. While ferocious and seemingly untouchable, army ants are at high risk from habitat fragmentation since they need large tracts of forest to support their enormous colonies (Gotwald 1995).

Leaf-cutting ants in the genus *Atta* also play a critical role in tropical forests. These ants can be considered "ecosystem engineers" since they have a large impact on their ecosystem and other organisms, primarily by moving soil as they create and maintain their large underground nest chambers.

Atta spp. cut leaves to grow a fungus garden (see below). While *Atta* spp. are considered a pest by many agricultural growers (including those in Kwamalasamutu) due to their tendency to cut the leaves of crops such as cassava, their role in the forest is irreplaceable. A single colony of *Atta sexdens* was documented to move 40 tons of soil to the surface in a forest in Brazil (Autori 1947). *Atta* structure the environment as they move soil, integrate nutrients, and aerate the soil for their large nests (Costa et al. 2008). In the rainforest, *Atta* do not tend to defoliate entire sections of forest due to the high density and diversity of tree species, from which they can selectively choose which leaves to cut. Crop monocultures are often hit hard when an *Atta* colony finds and cuts the plants. Planting a diverse selection of crops has been demonstrated to reduce the impacts of *Atta*. Other suggestions include 1) growing crops such as citrus (which *Atta* like) in another area to lure the *Atta* away from the main crops during growing season, and 2) mapping out the *Atta* colonies in an area before planting food crops. *Atta* spp. have very large, stationary nests that remain in the same place for many years (*Atta* queens can live for over 10 years). They also forage away from the nest on permanent trunk trails that are used for several years. Thus by mapping out the nests and trunk trails, gardens can be placed away from *Atta* foraging grounds. Poisons such as Mirex may work to kill an individual *Atta* nest, but they will also poison many other organisms and thus have a negative effect on the soil fauna. In addition, there are many sources of new *Atta* nests, such that poisoning them all is not possible (or advisable!).

Threats to the ants of the region and how to protect them

Like many tropical taxa, many ant species and populations face a range of threats. The most immediate and widespread threat comes from the loss, disturbance, or alteration of habitat. Fragmentation studies have revealed that ant species richness and genetic diversity can be affected even in large forest patches of 40 km² (Brühl et al. 2003, Bickel et al. 2006). Nomadic ant species such as army ants need large expanses of habitat to find enough food to feed their exceptionally large colonies (Gotwald 1995). Likewise, deforestation and forest fragmentation can cause local extinctions of the neotropical swarm-raiding army ant *Eciton burchellii* and other army ants (Boswell *et al.* 1998, Kumar and O'Donnell 2009).

Invasive ant species are a huge threat to native ant species. These aggressive species out-compete native ant species for food and other resources, or kill them directly, especially on islands and in degraded habitats. No invasive ant species were documented at the Werehpai RAP site but there were colonies of *Solenopsis geminata* and other ant species that proliferate in disturbed areas in the village of Kwamalasamutu and along the river near the RAP campsite. Care must be taken not to facilitate the spread of these ant species into the pristine forest.

Global climate change is likely already affecting the distribution of many ant species. For example, Colwell et al.

(2008) predict that as many as 80% of the ant species of a lowland rainforest could decline or disappear from the lowlands due to upslope range shifts and lowland extinctions (biotic attrition) resulting from the increased temperature of climate change.

A lack of information on ant species distributions, particularly for tropical regions such as Suriname, makes identifying rare and threatened species very difficult. Moreover, ants are small and easily overlooked by both the general public and conservationists, and are often perceived as pest organisms rather than focal species for conservation. While there are a few ant species that have become widespread invasive pests, most of the more than 12,000 described ant species are unobtrusive and beneficial to natural ecosystems and humans. Much conservation action is based on the assumption that other taxa, such as plants, birds or mammals, can serve as surrogates for the conservation needs of invertebrates and other lesser-known taxa (Rodrigues and Brooks 2007, Gardner et al. 2008). However, few studies or analyses of surrogacy have included ants; those that have generally indicate that ant diversity patterns and responses of ants to disturbance are not the same as those of most "umbrella taxa" (Alonso 2000). Ant species richness and distribution generally correlate best with other terrestrial, ground dwelling invertebrates (Alonso 2000) but these taxa are also not usually included in conservation planning.

Charismatic Ants

Given that ants are highly conspicuous and abundant around the Werehpai caves, they should be a key component of nature walks and eco-tourism visits to the site. Several ant species found in the Werehpai area are large enough to attract the attention and admiration of tourists. These ant species are common and conspicuous and have fascinating life histories and behaviors that give them "personalities" that tourists will find fascinating. Theses ant species can thus serve to highlight the key roles that ants play in the ecosystem. The most charismatic species around Werehpai include the following (see photos of these beautiful ants in the photo section):

Gigantiops destructor—the Jumping Ant—is a large black ant common on the forest floor in the Werehpai area. These ants have extremely large eyes with which to see and avoid their predators and their prey. They move very quickly and actually jump around on the leaf litter, which is unusual for an ant. Despite its name—*destructor*—these ants are timid, so you have to sneak up on them carefully. They do not bite or sting but defend themselves by spraying formic acid from their gaster (abdomen). These ants forage for small invertebrates in the leaf litter and are often found nesting near *Paraponera clavata* nests, possibly to benefit from the aggressive defense of the larger ants.

Daceton armigerum—the Canopy Ant—is a beautiful golden-colored ant that lives high in the canopy of trees near the Werehpai caves. They have large heads with strong muscles that power their sharp mandibles. Their eyes are under their head so that they can see below them as they walk along branches in the treetops. Another key to their success in the canopy is that their claws are very clingy and can keep a tight hold on branches and tree trunks.

Cephalotes atratus—the Turtle Ant, or Gliding Ant, lives high up in the tree canopy. With its flattened body and large turtle-shaped head, it lives within rotting twigs and branches and blocks the entrance to its nest with its head. Living so high in the canopy, these ants face the threat of falling out of their tree into the terrestrial territories of other, more ferocious ants. Thus they have evolved a way to avoid falling to the forest floor. If they fall from their tree, these ants stretch out their bodies and legs to glide (Yanoviak 2005). They can detect the tree trunk by the relative brightness against the dark greenery and twist in the air to point their abdomen toward their host tree, making a safe landing back home.

Eciton burchelli—the Army Ant—has very large colonies with millions of workers that move through the forest in a swarm raid, capturing everything in their path. These ants do not have a permanent nest but have a "bivouac"- a temporary nest site consisting of a giant ball of ants, usually found under a rotting log or in the hollow of a tree. These ants sting and bite and are very aggressive, even to humans, so one needs to watch where they step around these ants. It is very interesting to watch an army ant swarm since many other creatures can be seen jumping and running to get out of the path of the ants, and some specialized antbirds follow the swarm to catch these invertebrates for their meal. The soldiers of *E. burchelli* have very long mandibles which are used to suture wounds by some indigenous peoples. In addition to their swarms for catching food, these ants are also often seen moving their colony to a new bivouac (which is necessary when they run out of food in an area), carrying their larvae and pupae slung under their bodies.

Odontomachus spp.—Trap-Jaw Ants—are large ants common on the forest floor. These ants hold the world record for the fastest reflex in the animal kingdom. They forage by walking around with their mandibles (jaws) wide open. They have small trigger hairs between the mandibles which detect prey items (such as small invertebrates) and trigger the mandibles to snap shut very quickly to capture the prey. These ants often nest in the leaf litter trapped in small palm trees, in the terrestrial leaf litter, or in the soil. They are long, sleek, elegant ants, but have a nasty sting, so care must be taken to avoid touching them.

Paraponera clavata—the Bullet Ant or Congo Ant—is famous for its very powerful and painful sting. It is one of the world's largest ant species and is common in Neotropical lowland rainforests. These ants nest in the ground at the base of trees but forage up in the tree-tops on nectar and invertebrates. While they forage solitarily, they often have a relay of ants for passing large nectar droplets from the treetops to the nest, from one ant to another. These ants are one of the few ant species that make sound to communicate with one another. They can "stridulate" by rubbing their legs along their thorax to make a high-pitched squeaky sound.

Atta sp.—the Leaf-cutting Ant—is well known for its unique and fascinating agricultural lifestyle. *Atta* are fungus-growers- the workers cut pieces of leaves from a wide variety of trees to bring back to their nest where the leaves are chewed up by smaller workers and inserted into a large fungus garden, which the ants tend and cultivate. The ants do not feed on the leaves. Instead, they feed the fruiting bodies of the fungus to their larvae. Their nests are very large with many large underground chambers. It is fun to watch the workers cutting leaves and carrying them over their head back to the colony. *Atta* are parasitized by tiny phorid flies, which lay their eggs on the ants. When a fly larvae hatches, it burrows into an ant's head and develops inside, thereby killing the ant. Small *Atta* workers are often seen hitching a ride on the leaf carried by a larger worker- it is thought that these small ants serve to ward off attacking phorid flies.

Pseudomyrmex spp.— the Tree-dwelling Ants—are also arboreal ants. Many species live in the rotting, hollow twigs and branches up in the trees. They often fall from the trees, landing on the top of tents and even on your shirt, especially after a wind blows through the forest. Some species are specialized, obligate inhabitants of ant-plants, which provide a hollow cavity and sometimes food bodies or nectar for the ants. In exchange, the ants protect the plant by capturing and eating herbivorous insects that may eat the plant. These ants have large eyes and very long, slender bodies (their body form is distinctive) and a painful sting, so it's best to take care when observing them.

Ants and other invertebrates are an important part of the tropical ecosystem and must be considered in conservation and management planning (Alonso 2010). As E.O. Wilson (2006) has so aptly stated, "People need insects to survive, but insects do not need us. If all humankind were to disappear tomorrow, it is unlikely that a single insect species would go extinct, except three forms of human body and head lice…But if insects were to vanish, the terrestrial environment would soon collapse into chaos."

REFERENCES

Agosti, D., J.D. Majer, L.E. Alonso, T.R. Schultz (eds.). 2000. Ants: Standard Methods for Measuring and Monitoring Biological Diversity. Smithsonian Institution Press. Washington, D.C.

Alonso, L.E. (2000). Ants as indicators of diversity. In D. Agosti, J. Majer, L.E. Alonso and T.R. Schultz, eds. Ants, Standard Methods for Measuring and Monitoring Biodiversity, pp. 80–88. Smithsonian Institution Press, Washington, DC.

Alonso, L.E. 2010. Ant Conservation. *In:* L. Lach, C.L. Parr, and K.L. Abbott (editors), Ant Ecology, Oxford University Press, New York, USA.

Alonso, L.E. and Agosti, D. 2000. Biodiversity studies, monitoring, and ants: an overview *In:* Ants, Standard Methods for Measuring and Monitoring Biodiversity, D. Agosti, J. Majer, L. E. Alonso and T. R. Schultz (eds.). Washington, DC: Smithsonian Institution Press.

Autori, M. 1947. Combate a formiga saúva. Biologico. 13:196–199.

Bickel, T.O., Brühl, C.A., Gadau, J.R., Hölldobler, B., and Linsenmair, K.E. (2006). Influence of habitat fragmentation on the genetic variability in leaf litter ant populations in tropical rainforests of Sabah, Borneo. Biodiversity and Conservation. 15:157–175.

Borgmeier, T. 1934. Contribuição para o conhecimento da fauna mirmecológica dos cafezais de Paramaribo, Guiana Holandesa (Hym. Formicidae). Arquivos do Instituto de Biología Vegetal. 1:93–113.

Boswell, G.P., Britton, N.F. and Franks, N.R. (1998). Habitat fragmentation, percolation theory, and the conservation of a keystone species. Proceedings of the Royal Society of London B. 265:1921–1925.

Brühl, C.A., M. Mohamed, and K.E. Linsenmair. 1998. Altitudinal distribution of leaf litter ants along a transect in primary forests on Mount Kinabalu, Sabah, Malaysia. Journal of Tropical Ecology. 15: 265–177.

Brühl, C.A., Eltz, T., and Linsenmair, K.E. (2003). Size does matter—effects of tropical rainforest fragmentation on the leaf litter ant community in Sabah, Malaysia. Biodiversity and Conservation. 12:1371–1389.

Colwell, R., G. Brehm, C.L. Cardelús, A.C. Gilman, and J.T. Longino (2008). Global warming, elevational range shifts, and lowland biotic attrition in the wet tropics. Science. 322:258–261.

Costa, A.N., Vasconcelos, H.L.,Vieira-Neto, E.H.M., Bruna, E.M. (2008). Do herbivores exert top-down effects in Neotropical savannas? Estimates of biomass consumption by leaf-cutter ants (*Atta* spp.) in a Brazilian Cerrado site. *Journal of Vegetation Science.* 19:849–854.

Fernandez, F. and S. Sendoya. 2004. List of Neotropical ants (Hymenoptera: Formicidae). Biota Colombiana. 5: 3–93.

Fisher, B.L. 2005. A model for a global inventory of Ants: A case study in Madagascar. Proceedings of the California Academy of Sciences. 56: 86–97.

Fittkau, E.J. and H. Klinge. 1973. On biomass and trophic structure of the Central Amazonian rain forest ecosystem. Biotropica. 5:2–14.

Folgarait, P.J. (1998). Ant biodiversity and its relationship to ecosystem functioning: a review. Biodiversity and Conservation. 7:1221–44.

Gardner, T.A., Barlow, J., and Araujo, I.S. et al. (2008). The cost-effectiveness of biodiversity surveys in tropical forests. Ecology Letters. 11:139–150.

Gotwald, W. (1995). Army Ants: The Biology of Social Predation. Cornell University Press, Ithaca, NY, USA.

Kaspari, M. and J.D. Majer. 2000. Using ants to monitor environmental change. *In:* Ants, Standard Methods for Measuring and Monitoring Biodiversity, D. Agosti, J. Majer, L. E. Alonso and T. R. Schultz (eds.). Washington, DC: Smithsonian Institution Press.

Kempf, W.W. 1961. A Survey of the ants of the soil fauna in Surinam (Hymenoptera: Formicidae). Studia Entomologica. 4: 481–524.

Kempf, W.W. 1972. Catalogo abreviado das formigas da região Neotropical (Hymenoptera: Formicidae). Studia Entomologica. 15: 3–344.

Kumar, A. and O'Donnell, S. (2009). Elevation and forest clearing effects on foraging differ between surface—and subterranean—foraging army. Journal of Animal Ecology. 78:91–97.

LaPolla, J.S., T. Suman, J. Sosa-Calvo, and T.R. Schultz. 2007. Leaf litter ant diversity in Guyana. Biodiversity and Conservation. 16:491–510.

Longino, J.T., J. Coddington, and R.K. Colwell. 2002. The ant fauna of a tropical rain forest: Estimating species richness three different ways. Ecology 83: 689–702.

Lucky, A., E. Sarnat, and L.E. Alonso. 2011. Ants of the Muller Range, Papua New Guinea. In: Richards, S.J. and Gamui, B.G. (eds.) Rapid Biological Assessments of the Nakanai Mountains and the upper Strickland Basin: surveying the biodiversity of Papua New Guinea's sublime karst environments. RAP Bulletin of Biological Assessment 60. Conservation International, Arlington, VA.

Philpott, S.M., I Perfecto, I. Armbrecht, and C.L. Parr. 2010. Ant diversity and function in disturbed and changing habitats. *In:* L. Lach, C.L. Parr, and K.L. Abbott (editors), Ant Ecology, Oxford University Press, New York, USA.

Rodrigues, A.S.L. and T.M. Brooks, T.M. (2007). Shortcuts for biodiversity conservation planning: the effectiveness of surrogates. Annual Review of Ecology, Evolution and Systematics. 38:713–37.

Sosa-Calvo, J. 2007. Ants of the leaf litter of two plateaus in Eastern Suriname. *In:* Alonso, L.E. and J.H. Mol (eds). A Rapid Biological Assessment of the Lely and Nassau Plateaus, Suriname (with additional information on the Brownsberg Plateau). RAP Bulletin of Biological Assessment 43. Conservation International, Arlington, VA, USA.

Wilson, E.O. (2006). The Creation. W.W. Norton and Company. New York.

Yanoviak, S.P. and M. Kaspari. 2005. Directed aerial descent in canopy ants. Nature. 433: 624–6.

Appendix. A preliminary list of ant species of the Werehpai area, SW Suriname.

Ant morphospecies	ALL Transect near Werehpai caves	Hand collecting	Ant morphospecies	ALL Transect near Werehpai caves	Hand collecting
Acanthognathus lentis	X		*Hypoponera* 05		X
Acanthognathus 01		X	*Lachnomyrmex* 01	X	X
Apterostigma 01	X		*Myrmicocrypta guianensis*	X	
Atta 01 (*sexdens?*)		X	*Nesomyrmex* 01 (*spinodis?*)		X
Attini 01	X		*Nyleanderia* 01	X	
Attini 02	X		*Nyleanderia* 02	X	
Attini 03	X		*Nyleanderia* 03		X
Attini 04	X		*Octostruma balzani*		X
Attini 05	X		*Octostruma* 01	X	
Azteca 01	X		*Octostruma* 02	X	
Azteca 02		X	*Octostruma* 03	X	
Basicerotini 01	X		*Odontomachus* 01	X	
Basicerotini 02	X		*Odontomachus* 02	X	
Basicerotini 03	X		*Odontomachus* 03		X
Camponotus 01		X	*Odontomachus* 04	X	
Camponotus 02		X	*Odontomachus* 05	X	
Camponotus 03		X	*Pachycondyla* 01	X	
Camponotus 04		X	*Pachycondyla* 02	X	
Carebara nevermanni	X		*Pachycondyla* 03	X	
Cephalotes 01		X	*Pachycondyla* 04		X
Cephalotes atratus		X	*Pachycondyla* 05		X
Crematogaster 01	X		*Pachycondyla* 06		X
Crematogaster 02	X		*Paraponera clavata*		X
Crematogaster 03	X		*Pheidole* 01	X	
Crematogaster 04		X	*Pheidole* 02	X	
Crematogaster 05		X	*Pheidole* 03	X	
Crematogaster 06		X	*Pheidole* 04	X	
Crematogaster 07		X	*Pheidole* 05	X	
Cyphomyrmex laevigata	X		*Pheidole* 06	X	
Cyphomyrmex 01		X	*Pheidole* 07	X	
Daceton armigerum		X	*Pheidole* 08	X	
Discothyrea horni	X		*Pheidole* 09	X	
Dolichoderus 01		X	*Pheidole* 10	X	
Eciton burchelli		X	*Pheidole* 11	X	
Eciton 01		X	*Pheidole* 13	X	
Gigantiops destructor		X	*Pheidole* 14	X	
Gnamptogenys minuta	X		*Pheidole* 15		X
Gnamptogenys 01		X	*Pheidole* 16		X
Hypoponera 01	X		*Pheidole* 17		X
Hypoponera 02	X		*Pheidole* 18		X
Hypoponera 03	X		*Pheidole* 19		X
Hypoponera 04	X		*Pheidole* 20		X

table continued on next page

Ant morphospecies	ALL Transect near Werehpai caves	Hand collecting
Pseudomyrmex 01		X
Pseudomyrmex 02		X
Pseudomyrmex 03		X
Pyramica denticulata	X	
Pyramica subdentata	X	
Pyramica alberti	X	
Pyramica glenognatha	X	
Pyramica 01		
Solenopsis 01	X	
Solenopsis 02	X	
Solenopsis 03	X	
Solenopsis 04	X	
Solenopsis 05	X	
Solenopsis 06		X
Solenopsis 07		X
Strumigenys elongata		X
Strumigenys 01	X	
Strumigenys 02	X	
Trachymyrmex ruthae (species group)	X	
Wasmannia auropunctata	X	X
Total number of species (105)	62	44

Chapter 8

Fishes of the Sipaliwini and Kutari Rivers, Suriname

Philip W. Willink, Kenneth Wan Tong You, and Martino Piqué

SUMMARY

Forty-three sites near three camps along the Sipaliwini and Kutari rivers, Suriname were sampled between August 19 and September 5, 2010. We recorded 99 species of fishes. This diversity is high compared to the rest of the world, but is typical for the Guiana Shield. We collected eight species of fishes potentially new to science, including a large catfish with spines along the body and a small catfish that lives in sand-bottomed creeks. Two species are new records for Suriname. We collected 57 species at Kutari (Site 1), 60 species at Sipaliwini (Site 2), and 63 species at Werehpai (Site 3). This is remarkably consistent, with no significant difference in diversity among camps. However, we did not necessarily find the same species at each camp. Creek assemblages were similar among the three sites. Many young fishes were found in flooded forests, even if the adults lived in rivers or other habitats. Overall, large top-level predators were uncommon. The region is exhibiting the first stages of overfishing. Many fishes still occur in the Sipaliwini area, but there is a need to assess fishing pressure and implement management plans.

INTRODUCTION

Fishes are a critical source of protein in the Kwamalasamutu region. They are a common component of many meals. To further emphasize the importance of fishes to the people living in the area, some regional geographic names are even based on local fish. For example, 'sipali' means stingray and 'wini' means river/water in Carib, of which the Trio language is part. In other words, Sipaliwini can be translated 'river of stingrays' (Boven 2006).

Despite the importance of fishes in the indigenous culture, relatively few species are routinely eaten. Most species are small and often ignored by people, but they are actually an important part of the aquatic ecosystem. Smaller fishes forage on aquatic insects and serve as prey for larger fishes, caiman, and birds. Fish diversity reflects the health of the river systems.

There is a modest amount of published information concerning the fishes of the region. Most fish surveys in the watershed are well to the north in the Corantijn proper (e.g., Vari 1982). Some fish collections have been made near regional airstrips, such as those at Kwamalasamutu and the Sipaliwini Savanna. These specimens were then used as the basis for species descriptions (e.g. Gery 1961) or reviews of specific taxonomic groups (e.g. cichlids by Kullander and Nijssen 1989). Ouboter and Mol (1993) reported 75 species in the Curuni/Sipaliwini basin, a much larger area than was surveyed during this expedition. Any other information is unpublished. To our knowledge, there have been no prior scientific fish collections in the Kutari River or Wioemi Creek. This expedition was the first, and will serve as a baseline for subsequent aquatic biodiversity studies.

STUDY SITES AND METHODS

Forty-three sites near the three camps were sampled between August 19 and September 5, 2010. Fishes were collected with a 3-meter fine-mesh seine, 5-meter seine, dip nets, 30-meter trammel net, 40-meter experimental gillnet, and hook and line. We also talked extensively with our Trio guides about their knowledge of local fishes, and to discern what they were catching during the expedition. Every habitat was sampled with as many methods as practical in order to rapidly assess the diversity of the region and maximize the number of species observations. Rocks in rapids were scraped. Submerged logs were cut open. Leaf litter was searched. Seines were pulled through patches of vegetation, as well as over sandy beaches. Dip nets were dragged through flooded tree branches. Canoes were used to travel extensively upstream and downstream from the camps. We also walked through the forest to survey creeks and swamps. Most individuals were released, but representative specimens were preserved in 4% formalin and later transferred to 70% ethanol for long-term storage at the National Zoological Collection of Suriname in Paramaribo and The Field Museum in Chicago, USA.

The Kutari River at Site 1 was approximately 40 meters wide and meandered extensively. There was a significant flood plain, and much of the vegetation along the river was submerged during the time of the expedition. No rocks, rapids, or beaches were apparent due to the high water levels. Current was fast flowing. Creeks were usually sampled well inside the forest and distant from the main channel of the Kutari River. No people were seen, but there were scattered abandoned campsites along the river.

The Sipaliwini River at Site 2 was approximately 75 meters wide and the primary river channel was relatively straight. Large boulders were common, and rapids were present, although most were still submerged during the time of the expedition. Aquatic plants grew on rocks in the rapids. Islands and sand beaches were beginning to emerge as the river level dropped. A few people were observed fishing, using gillnets and hook and line. Creeks were usually sampled near their confluence with the Sipaliwini River.

The Sipaliwini River at Site 3 (Werehpai), downstream from Site 2, was very similar. The river was larger at this site, approximately 150 meters wide, and eddies and bays were also larger. There were several adjacent swamps. The morphology of the creeks in the forest was similar to the other two sites. Wioemi Creek is better described as a small river, almost as large as the Kutari and very similar geomorphologically. Many people were observed fishing in this area, since it is closer than the other two sites to the town of Kwamalasamutu.

RESULTS AND DISCUSSION

We recorded 99 species of fishes (Appendix). This is typical for the interior of Suriname and nearby parts of the Guiana Shield; for example, the Coppename RAP recorded 112 species (Mol et al. 2006), the Eastern Kanuku Mountains RAP recorded 113 species (Mol 2002), and a similar rapid assessment of the upper Essequibo River yielded 110 species (P.W. Willink et al. *unpublished data*). This diversity is high compared to the rest of the world, but is typical for the Guiana Shield.

The species accumulation curve for the expedition is showing signs of reaching an asymptote (Fig. 1). After collection station #23, far fewer novel species were recorded during the remainder of the survey. Exceptions were stations #33–34 that were in rapids. We had not surveyed many rapids prior to these stations, so they added several species to our cumulative list. But after this point, we had sampled essentially all available habitats in the area under study. There are still many species in the region that we probably did not collect due to the high water and seasonal effects. More surveys need to be done at different times of the year.

We collected eight species of fishes potentially new to science, including a large catfish *Pterodoras* aff. *granulosus* with spines along the body and a small catfish *Imparfinis* aff. *stictonotus* that lives in sand-bottomed creeks. The other potentially new species are *Pseudacanthicus* sp., *Hypostomus* aff. *taphorni*, *Eigenmannia* sp. 1, *Eigenmannia* sp. 2, *Astyanax* sp., and *Moenkhausia* aff. *georgiae*. The two *Eigenmannia* species were initially recognized on a previous RAP in the Coppename River (Willink and Sidlauskas 2006), and appear to be more widely distributed throughout Suriname's interior than originally thought. This number of new species is typical for Neotropical rivers that have not been well surveyed by fish biologists.

Two species, *Ituglanis gracilior* and *Hemigrammus orthus*, were new records for the country of Suriname. They were previously only found in Guyana to the west (Vari et al. 2009). We encountered some species of fishes known only from the Sipaliwini River and nearby drainages (e.g., *Corydoras sipaliwini* and *Crenicichla sipaliwini*). Other species, such

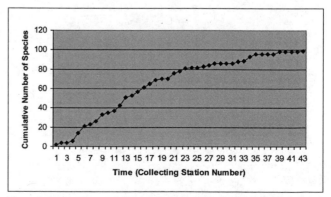

Figure 1. Species accumulation curve for fishes collected in the Kutari and Sipaliwini Rivers, Suriname, August 19 to September 5, 2010.

as *Moenkhausia collettii* and *Hoplias malabaricus*, are widespread throughout the Guianas and much of South America.

We collected 57 species at Kutari (Site 1), 60 species at Sipaliwini (Site 2), and 63 species at Werehpai (Site 3) (Appendix). This is remarkably consistent, with no significant difference in diversity among sites. However, we did not necessarily find the same species at each site. For example, *Hoplias aimara* was most common in the meandering flooded Kutari River. Knifefish diversity was higher in the Kutari as well. In comparison, at Site 2 the Sipaliwini River had more rapids and flowing water, and this was even more the case at Site 3. Piranhas and larger catfishes were more common at these sites. Tucunaré (*Cichla ocellaris*) and large characids were also present.

Wioemi Creek was similar in geomorphology to the Kutari River. Both had numerous meanders and S-curves in the main channel. Banks were relatively low and generally covered with flooded shrubs because of the high water levels. Large sections of the adjacent forest were flooded as well.

Wioemi Creek and the Kutari River are also similar in species composition. We only sampled four localities in Wioemi Creek, and recorded 23 species. This is not rigorous enough to make any meaningful statistical comparisons. But of the 23 species, 16 were shared with Kutari, 13 with Sipaliwini, and 8 with Werephai. Also, six species (*Hemigrammus orthus*, *Hyphessobrycon rosaceus*, *Odontostilbe gracilis*, *Phenacogaster carteri*, *Melanocharacidium dispilomma*, and *Otocinclus mariae*) were found exclusively in the Kutari River and Wioemi Creek (but not in the Sipaliwini River at Sites 2 or 3). This indicates that habitat plays an important role in species distribution.

The role of habitat was apparent throughout the entire region. Particular species were found in particular habitats, regardless of which sub-basin they were in. For example, creeks were characterized by *Pyrhulina stoli*, *Jupiaba abramoides*, *Rineloricaria*, and *Rivulus*. Rapids were characterized by *Pseudancistrus corantijniensis*, *Lithoxus* aff. *bovallii*, and *Guayanacistrus brevispinis*. Larger and deeper sections of the rivers held *Schizodon fasciatus*, *Hemisorubim platyrhynchos*, and *Prochilodus rubrotaeniatus*.

During this survey, many young fishes were found in flooded forests, even if the adults lived in rivers or other habitats. This is because many fishes spawn in flooded areas at the beginning of the rainy season, which was several months prior to the expedition. This strategy gives young fishes an opportunity to find hiding places among submerged vegetation. There is also an increased amount of food in the form of insects falling into the water, suspension of nutrients from the leaf litter, greater access to seeds and fruits, etc. (Roberts 1972, Goulding 1980, Lowe-McConnell 1987). The RAP survey was conducted at the end of the rainy season, so we found many fishes that were only a few months old. This demonstrates the importance of seasonal flooding and the interconnection of terrestrial and aquatic habitats. If anything negatively impacts the forest, it will also impact the fishes in the river.

We recorded a small number of very large piranhas around Sipaliwini (Site 2), including a black piranha *Serrasalmus rhombeus* that was 41 centimeters in standard length and weighed 3 kilograms (see color plates). We found numerous piranhas nearer to Kwamalasamutu, but they were almost all juveniles or small adults. Large catfishes were rare, as were tucunaré. Small tetras were abundant at Sipaliwini and Werehpai, but far less so in the Kutari River (where the large predator *Hoplias aimara* was most common). Usually there are fewer small tetras in areas with many predators. This is consistent with what we observed. Overall, large top-level predators were uncommon. In pristine environments, these types of fishes are abundant, but they are the first to disappear when there is excessive fishing pressure (Mol et al. 2006). We often saw people fishing along the river, and nearly every household had a gill net. The region is exhibiting the first stages of overfishing. Many fishes still occur in the Sipaliwini area, but there is a need to assess fishing pressure and implement management plans.

The primary threat to the fishes of the Sipaliwini River is overfishing. Fishes are an important source of protein in the region, and people in Kwamalasamutu have to travel hours from the village in order to find large fishes. Fish diversity is still high, but popular food fishes are decreasing in size, and some are becoming less common (e.g., red-tailed catfish *Phractocephalus hemioliopterus*). Logging would have negative impacts by increasing erosion and decreasing the amount of food that falls into the water, especially when the rivers flood. We are unaware of any imminent plans to deforest the region. We are also unaware of any plans for gold or bauxite mining. However, diamond exploration concessions exist in a watershed well upstream. Excessive mining would result in erosion and sedimentation, negatively impacting fishes, especially those that live along the bottom.

CONSERVATION RECOMMENDATIONS

- Assess which fish species are used for food. Determine amount caught and eaten. Study life-history of these species to determine how fast they reproduce and grow.
- Determine the amount of fish that can be sustainably harvested. Set catch limits and/or seasons if necessary to avoid overfishing.
- Create picture guides of fishes, especially colorful species and fun-to-catch species, in order to increase appreciation and knowledge of fishes. These guides can be used to promote ecotourism.
- Maintain forests along rivers, especially in areas that flood. This is to prevent erosion and maintain the amount of nutrients (i.e., insects, leaves, fruits, etc.) that fall into the water and act as fish food. Flooded areas, such as Kutari River and Wioemi Creek, are important breeding grounds for fishes.

- Additional scientific surveys are necessary to document the fish biodiversity. There are species present that we did not collect, and there could be new species to science yet to be discovered. Additional surveys should be conducted at different times of the year, especially when river levels are lower. These surveys could also explore further upstream and downstream than we traveled.

LITERATURE CITED

Boven, K.M. 2006. Overleven in een grensgebied: veranderingsprocessen bij de Wayana in Suriname en Frans Guyana. Bronnen voor de studie van Suriname. Deel 26. IBS/Rozenberg Publishers, Leiden and Amsterdam.

Gery, J.R. 1961. *Hyphessobrycon georgetti* sp. nov., a dwarf species from southern Suriname. Bulletin of Aquatic Biology 2:121–128.

Goulding, M. 1980. The fishes and the forest. University of California Press, Berkley.

Kullander, S.O. and H. Nijssen. 1989. The cichlids of Suriname. E.J. Brill, Leiden, Netherlands.

Lowe-McConnell, R.H. 1987. Ecological studies in tropical fish communities. Cambridge University Press, Cambridge.

Mol, J.H. 2002. A preliminary assessment of the fish fauna and water quality of the Eastern Kanuku Mountains: Lower Kwitaro River and Rewa River at Corona Falls. pp. 38–42. in: Montambault, J.R. and O. Missa (eds.). A Biodiversity Assessment of the Eastern Kanuku Mountains, Lower Kwitaro River, Guyana. RAP Bulletin of Biological Assessment 26. Washington, DC: Conservation International.

Mol, J.H., P. Willink, B. Chernoff, and M. Cooperman. 2006. Fishes of the Coppename River, Central Suriname Nature Reserve, Suriname. pp. 67–79 *in*: Alonso, L.E., and H.J. Berrenstein (eds). A Rapid Biological Assessment of the Aquatic Ecosystems of the Coppename River Basin, Suriname. RAP Bulletin of Biological Assessment 39. Washington, DC: Conservation International.

Ouboter, P.E. and J.H.A. Mol. 1993. The fish fauna of Suriname. Pp. 133–154 in: Ouboter, P.E. (ed.). Freshwater Ecosystems of Suriname. Kluwer Academic Publishers, Netherlands.

Roberts, T.R. 1972. Ecology of fishes in the Amazon and Congo basins. Bulletin of the Museum of Comparative Zoology 143:117–147.

Vari, R.P. 1982. Environmental impact of the Kabalebo Projekt. Inventory, biology, and ecology of the fishes in the Corantijn River system, Suriname. Unpublished report to the Ministry of Development, Republic of Suriname.

Vari, R.P., C.J. Ferraris, Jr., A. Radosavljevic, and V.A. Funk (eds.). 2009. Checklist of the freshwater fishes of the Guiana Shield. Bulletin of the Biological Society of Washington 17:1–95.

Willink, P.W. and B.L. Sidlauskas. 2006. Taxonomic notes on select fishes collected during the 2004 AquaRAP expedition to the Coppename River, Central Suriname Nature Reserve, Suriname. pp. 101–111 *in*: Alonso, L.E., and H.J. Berrenstein (eds). A Rapid Biological Assessment of the Aquatic Ecosystems of the Coppename River Basin, Suriname. RAP Bulletin of Biological Assessment 39. Washington, DC: Conservation International.

Appendix. List of fishes recorded in the Kutari and Sipaliwini rivers.

Taxon	Kutari	Sipaliwini	Werehpai
RAJIFORMES			
Potamotrygonidae			
Potamotrygon boesemani	X	X	
CHARACIFORMES			
Acestrorhynchidae			
Acestrorhynchus microlepis	X	X	X
Anostomidae			
Anostomus anostomus	X		X
Hypomasticus megalepis			X
Leporinus fasciatus			X
Leporinus friderici	X		
Leporinus granti		X	X
Leporinus nijsseni	X		
Schizodon fasciatus	X		X
Characidae			
Astyanax bimaculatus		X	X
Astyanax sp.		X	
Brachychalcinus orbicularis	X	X	X
Brycon falcatus	X	X	X
Bryconamericus hyphesson		X	X
Bryconops affinis	X	X	X
Bryconops melanurus		X	
Chalceus macrolepidotus	X		X
Charax gibbosus	X	X	X
Cynopotamus essequibensis	X	X	X
Hemigrammus ocellifer	X	X	
Hemigrammus orthus	X		X
Hyphessobrycon rosaceus	X		X
Jupiaba abramoides	X	X	X
Jupiaba meunieri		X	X
Jupiaba polylepis		X	X
Moenkhausia chrysargyrea	X	X	
Moenkhausia collettii	X	X	X
Moenkhausia georgiae	X	X	X
Moenkhausia grandisquamis			X
Moenkhausia hemigrammoides		X	X
Moenkhausia lepidura	X	X	X
Moenkhausia oligolepis	X	X	X
Myleus rhomboidalis	X	X	X
Myloplus rubripinnis	X	X	X
Odontostilbe gracilis	X		X
Phenacogaster carteri	X		X

Taxon	Kutari	Sipaliwini	Werehpai
Poptella longipinnis		X	X
Roeboexodon guyanensis	X		X
Serrasalmus rhombeus	X	X	X
Tetragonopterus chalceus		X	X
Tetragonopterus rarus	X		
Triportheus brachipomus		X	X
Crenuchidae			
Characidium zebra	X	X	X
Melanocharacidium dispilomma	X		X
Curimatidae			
Cyphocharax helleri	X	X	
Cyphocharax spilurus	X	X	X
Cynodontidae			
Cynodon gibbus	X	X	X
Erythrinidae			
Hoplerythrinus unitaeniatus		X	
Hoplias aimara	X	X	X
Hoplias curupira	X	X	
Hoplias malabaricus	X		
Gasteropelecidae			
Carnegiella strigata	X	X	X
Hemiodontidae			
Bivibranchia bimaculata		X	
Hemiodus argenteus		X	
Hemiodus quadrimaculatus			X
Lebiasinidae			
Pyrrhulina stoli	X	X	X
Parodontidae			
Parodon guyanensis			X
Prochilodontidae			
Prochilodus rubrotaeniatus		X	X
SILURIFORMES			
Auchenipteridae			
Ageneiosus inermis	X	X	X
Tatia intermedia	X		
Trachelyopterus galeatus		X	
Callichthyidae			
Corydoras baderi	X		
Corydoras sipaliwini		X	X
Cetopsidae			
Cetopsidium minutum	X		
Helogenes marmoratus	X		

table continued on next page

Taxon	Kutari	Sipaliwini	Werehpai
Doradidae			
Doras carinatus		X	X
Pterodoras aff. *granulosus*		X	
Heptapteridae			
Imparfinis aff. *stictonotus*	X		X
Pimelodella cristata	X		X
Pimelodella macturki		X	
Loricariidae			
Ancistrus aff. *leucostictus*		X	
Cteniloricaria platystoma		X	X
Guyanancistrus brevispinis			X
Hypostomus pseudohemiurus	X	X	X
Hypostomus taphorni			X
Lithoxus aff. *bovallii*			X
Metaloricaria nijsseni		X	
Otocinclus mariae	X		X
Pseudacanthicus sp.		X	
Pseudancistrus corantijniensis		X	X
Rineloricaria sp.	X		
Rineloricaria stewarti			X
Pimelodidae			
Hemisorubim platyrhynchos			X
Pimelodus blochii		X	
Pimelodus ornatus		X	
Pseudopimelodidae			
Microglanis secundus		X	X
Trichomycteridae			
Ituglanis gracilior	X		
GYMNOTIFORMES			
Gymnotidae			
Gymnotus carapo	X		X
Hypopomidae			
Brachyhypopomus brevirostris	X		
Rhamphichthyidae			
Rhamphichthys rostratus			X
Sternopygidae			
Eigenmannia sp. 1	X		
Eigenmannia sp. 2	X		
CYPRINODONTIFORMES			
Rivulidae			
Rivulus sp.	X	X	

Taxon	Kutari	Sipaliwini	Werehpai
SYNBRANCHIFORMES			
Synbranchidae			
Synbranchus marmoratus	X		
PERCIFORMES			
Cichlidae			
Apistogramma steindachneri	X		X
Cichla ocellaris		X	X
Crenicichla sipaliwini	X	X	
Geophagus brachybranchus		X	
Guianacara sphenozona		X	
Total 99 species	**57**	**60**	**63**

Chapter 9

A rapid assessment of the amphibians and reptiles of the Kwamalasamutu region (Kutari/lower Sipaliwini Rivers), Suriname

Paul E. Ouboter, Rawien Jairam, and Cindyrella Kasanpawiro

SUMMARY

The RAP team recorded 42 species of amphibians and 36 species of reptiles, including one species of frog in the genus *Hypsiboas* that is new to science. The amphibian community was most similar to those of forests on bauxite plateaus in western Suriname. Some rare species were collected, of which the tree frog *Osteocephalus cabrerai* and the amphisbaenian *Amphisbaena slevini* were collected from Suriname for the first time. We also encountered *Chelonoides denticulata* (Yellow-footed Tortoise), listed as Vulnerable on the IUCN Red List. Apart from caimans, most of the herpetofauna of the area seems to be minimally impacted by human activities such as hunting and fishing from the community of Kwamalasamutu. We discovered that certain expected species that are quite common in other areas in Suriname were either not found or found in very moderate numbers on the RAP survey. On the other hand, we found certain generally rare species to be quite common, emphasizing the importance of the region's forests to the biodiversity of Suriname and the Guiana Shield. Recommended conservation measures include avoiding large-scale deforestation in the region, and controlling the hunting of caimans.

INTRODUCTION

Amphibians are very important indicators of disturbance, because they are sensitive to changes in microclimate, and worldwide many species recently became extinct as the result of several impacts, including habitat change, pollution and disease. The group is well suited for rapid assessments, as the species are often easy to sample and their calls are diagnostic, which aids identification of species that cannot be collected, such as tree frogs in the rainforest canopy. Of the reptiles, lizards are often sensitive to changes in microhabitat. Many species are restricted to pristine forest habitats, and tend to disappear from regions of highly degraded forest. Caimans, turtles, and tortoises are generally good indicators of hunting pressure.

Although two previous expeditions by the National Zoological Collection / Anton de Kom University of Suriname visited southern Suriname in 1988 and 1989, these expeditions focused on the Sipaliwini Savanna and the Apalagadi area north of the savanna, and did not survey any areas in the vicinity of Kwamalasamutu. The Kutari River has never been visited by a biological expedition before. The goal of this survey was to provide baseline information on the diversity and abundance of reptiles and amphibians in the forests of the Kwamalasamutu region.

METHODS

Four areas along the Kutari and Sipaliwini Rivers were investigated for amphibians and reptiles. Six days were spent at each of the three RAP sites (see Executive Summary for site descriptions). Iwana Samu was visited for only 2 days, but collections were made there as well.

We sampled the herpetofauna by walking trails at each survey site and searching for animals along the trails. Most of the trails had been cut during the month before the RAP survey, and some trails were extended during our stay at each site. Trails were walked during morning, early afternoon and evening to observe and/or collect amphibians and reptiles. Special attention was given to creeks, downed logs, cavities, and other favourable habitats, to discover as many species of amphibians and reptiles as possible.

Specimens were captured by hand. Frog calls were recorded using a Marantz PMD660 solid state recorder and Sennheiser ME 67 directional microphone. Calls were compared with known calls for the frogs of French Guiana (Marty & Gaucher 2000) and Ecuador (Read 2000).

Many specimens were identified on site. Some specimens, particularly those that could not be identified conclusively in the field, were collected and preserved for later identification in the laboratory. Specimens were euthanized using an injection of Nembutal, fixed in 4% formaldehyde solution, and subsequently preserved in 70% ethanol. All specimens are deposited in the National Zoological Collection of Suriname.

Species diversity and richness were calculated using Simpson's index of diversity, the Shannon-Wiener diversity index, and Chao 1 (Magurran 2004), using the software package Species Diversity & Richness IV (Pisces Conservation Ltd). PCA analysis was accomplished with the software Community Analysis Package 4 of the same provider.

RESULTS AND DISCUSSION

A total of 42 species of amphibians and 36 species of reptiles was observed in the area (Appendix). An estimation of the total number of species in the area based on Chao & Lee 1 is 42.84 amphibians and 43.34 reptiles. It can therefore be concluded that the sampling of amphibians was adequate, but that continued sampling of reptiles would probably yield additional species.

The most exciting discovery was a tree frog in the genus *Hypsiboas* that is new to science. This frog was discovered in swamp forest directly adjacent to the camp at the Kutari River site; one specimen was collected. Photos of this frog, as well as descriptive notes, will appear in the forthcoming Amphibians of Suriname (Ouboter and Jairam *in press*).

The diversity of the areas investigated is shown in Table 1. The Sipaliwini site had the highest species diversity of the three RAP sites. Werehpai had the greatest number of reptile species, but high abundance of three species of lizards significantly decreased evenness at the site, and consequently the α diversity values.

The PCA analysis showed that the community structure of the three RAP sites differed significantly, and that the findings from all three (or even more) sites are needed to obtain an overview of the herpetofauna of the area. A comparison of the amphibian community of the Kwamalasamutu area with other areas of Suriname is shown in Fig. 1. Within Suriname, the amphibian community of the Kwamalasamutu region is most similar to the communities on forests on bauxite plateaus, but is probably slightly poorer in the number of species, especially in the families Aromobatidae and Caeciliidae.

A number of rare species was collected during the survey: *Osteocephalus cabrerai*, a rare tree frog from the western Amazon Basin and French Guiana, is herewith reported from Suriname for the first time. *Scinax proboscideus* is a tree frog with a nasal appendix, previously known from only two localities in the interior of Suriname and a few localities in French Guiana (Ouboter and Jairam *in press*). *Microcaecilia taylori* was described from three specimens collected from forest islands in the Sipaliwini Savanna (Nussbaum

Table 1. Alpha diversity of the three RAP sites. It should be noted that the time spent at Iwana Samu was much less than at the other three sites.

	Site 1 Kutari	Site 2 Sipaliwini	Site 3 Werehpai	Iwana Samu
Amphibians				
Species richness	23	27	26	12
Simpson's index	13.86	15.40	11.38	7.02
Shannon-Wiener	2.73	2.92	2.75	2.07
Reptiles				
Species richness	14	13	21	1
Simpson's index	6.56	17.10	7.55	-
Shannon-Wiener	2.13	2.41	2.38	-

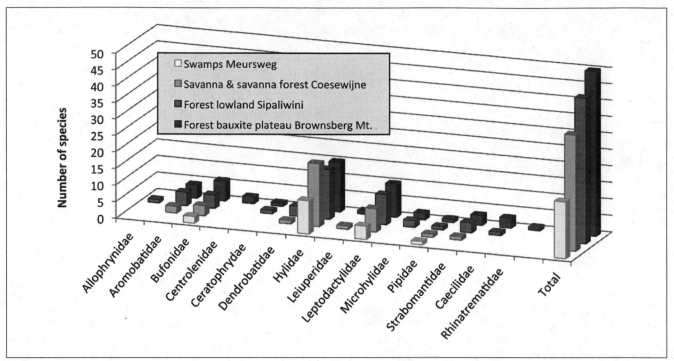

Figure 1. Comparison of the amphibian community of the Kwamalasamutu region (lowland forest) with amphibian communities at Brownsberg (forest on bauxite plateaus), Boven Coesewijne area (savanna & savanna forest) and the Meursweg area (freshwater swamps).

& Hoogmoed 1979). The specimen collected by us is the fourth, and shows that this species is not restricted to the Sipaliwini Savanna. The snake *Xenodon werneri* is quite rare and was previously known from only two specimens in Suriname, one from the Wilhelmina Mts. and the other from the Nassau Mts. (Hoogmoed 1985). The amphisbaenian *Amphisbaena slevini*, known from the surroundings of Manaus (Brasil) and eastern French Guiana (Starace 1998), was collected on the RAP, providing the first record for Suriname and an extension of its known range.

Several species observed on the RAP survey are indicators of relatively undisturbed forests, including *Allobates granti*, *Ceratophrys cornuta*, *Ameerega hahneli*, *Hypsiboas fasciata*, *Leptodactylus hyeri*, *Chiasmocleis shudikarensis*, *Pristimantis chiastonotus*, *Pristimantis marmoratus*, *Bothriopsis biliniatus*, and *Gonatodes annularis*. Typical indicator species of forest clearing were absent. A generally anthropogenic species, *Rhinella marina*, was present at the Werehpai site.

We observed several species of amphibians and reptiles listed in Appendix II of CITES, including species in the genera *Allobates*, *Ameerega*, *Dendrobates*, *Tupinambis* and *Paleosuchus*, and species of the families Boidae and Testudinidae. However, almost all of these are listed as "Least Concern" in the IUCN Red Data List (see Appendix). Exceptions are *Chelonoides denticulata*, which is listed as "Vulnerable", and *Paleosuchus trigonatus*, which is listed as "Lower Risk/Least Concern (needs updating)".

Most amphibians and reptiles are opportunistic predators of small to medium size animals: arthropods for frogs and lizards; annelids and the like for amphisbaenians; caecilians,

small snakes, and a variety of small vertebrates for snakes, turtles and caimans. Because of their opportunistic feeding behavior, their role in the ecosystem per species is probably minor. An exception is the role of caimans as top predators in the aquatic ecosystem (Fittkau 1973). A healthy caiman population may positively influence fish stocks and other characteristics of wetland ecosystems.

CONSERVATION RECOMMENDATIONS

In general, the ecosystems investigated seem to have a healthy and diverse herpetofauna. The current intensity of fishing, hunting, and gathering activities does not seem to adversely affect most amphibian and reptile populations in the Kwamalasamutu region. Such activities are expected to have more impact if extensive deforestation occurs; therefore, large-scale deforestation should be prevented at all costs.

Although we did not carry out a targeted survey for caimans, the impression is the same as in 1988 (Ouboter 1989): near Kwamalasamutu, caimans are over-exploited and therefore rare. Every caiman encountered away from the village, e.g. in the Kutari area, is instantly killed by the local inhabitants. Caimans have a positive effect on fish populations and community structure, and should therefore be provided some form of protection in the vicinity of Kwamalasamutu and the broader region. This could be accomplished in several ways, including a no-hunting agreement during part of the year or during alternate years.

REFERENCES

Avila-Pires, T.C.S. 1995. Lizards of Brazilian Amazonia. Zool. Verh. 299: 1–706.

Fittkau, E.J. 1973. Crocodiles and the nutrient metabolism of Amazonian waters. Amazoniana 4(1): 103–133.

Hoogmoed, M.S. 1973. Notes on the herpetofauna of Surinam IV. The lizards and amphisbaenians of Surinam. Junk, The Hague, 419 pp.

Hoogmoed, M.S. 1985. *Xenodon werneri* Eiselt, a poorly known snake from Guiana, with notes on *Waglerophis merremii* (Wagner) (Reptilia: Serpentes: Colubridae). Notes on the herpetofauna of Surinam IX. Zool. Meded. Leiden 59(8): 79–88.

Lescure, J. & C. Marty, 2000. Atlas des Amphibiens de Guyane. M.N.H.N., Paris, 388 pp.

Magurran, A.E. 2004. Measuring Biological Diversity. Blackwell Publ., Maiden, 256 pp.

Marty, C. & P. Gaucher, 2000. Sound guide to The Tailless Amphibians of French Guiana.

Nussbaum, R.A. & M.S. Hoogmoed. 1979. Surinam Caecilians, with notes on *Rhinatrema bivittatum* and the description of a new species of *Microcaecilia* (Amphibia, Gymnophiona). Zool. Meded. Leiden 54(14): 217–235.

Ouboter, P.E. 1989. The Impact of an Indian Village on Caimans. (Area Reports: Suriname). IUCN Crocodile Specialist Group Newsletter 8: 28.

Ouboter, P.E. 1996. Ecological studies on crocodilians in Suriname – niche segregation and competition in three predators. SPB Acad. Publ., Amsterdam, 139 pp.

Ouboter, P.E. & R. Jairam, in press. Amphibians of Suriname. E.J. Brill, Leiden.

Read, M. 2000. Frogs of the Ecuadorian Amazon – a guide to their calls.

Starace, F. 1998. Guide des Serpents et Amphisbenes de Guyane. Ibis Rouge, Guadeloupe/Guyane, 449 pp.

Appendix. List of amphibians and reptiles found during the Kwamalasamutu RAP survey. Numbers indicate number of observations per site; they do not necessarily indicate that specimens were collected.

CLASS/(Sub)Order/Family	Species	Indicator of pristine forest	IUCN Red List Status	Kutari	Sipaliwini	Werehpai	Iwana Samu
AMPHIBIA							
ANURA							
Allophrynidae	*Allophryne ruthveni*		Least concern	4	2	4	1
Aromatidae	*Allobates femoralis*		Least concern	1	2	34	2
	Allobates granti	X	Least concern	6	0	4	0
	Anomaloglossus baeobatrachus		Data deficient	13	7	7	0
Bufonidae	*Rhaebo guttatus*		Least concern	3	2	5	1
	Rhinella martyi		Least concern	3	0	0	0
	Rhinella lescurei		Data deficient	12	13	9	0
	Rhinella marina		Least concern	0	0	1	0
Ceratopryidae	*Ceratophrys cornuta*	X	Least concern	0	0	2	0
Dendrobatidae	*Dendrobates tinctorius*		Least concern	0	5	0	0
	Ameerega hahneli	X	Least concern	1	1	10	0
	Ameerega trivittata		Least concern	8	2	11	0
Hylidae	*Dendropsophus minutus*		Least concern	0	4	0	0
	Hypsiboas boans		Least concern	5	14	1	1
	Hypsiboas calcaratus		Least concern	1	22	0	6
	Hypsiboas cinerascens		Least concern	2	1	8	0
	Hypsiboas fasciatus	X	Least concern	0	3	6	0
	Hypsiboas geographicus		Least concern	0	4	0	0
	Hypsiboas sp. nov.		Unknown	1	0	0	0
	Osteocephalus buckleyi		Least concern	0	0	0	1
	Osteocephalus cabrerai		Least concern	0	1	0	1
	Osteocephalus leprieuri		Least concern	3	2	0	0
	Osteocephalus taurinus		Least concern	0	0	1	0
	Scinax proboscideus		Least concern	1	0	0	0
	Trachycephalus resinifictrix		Least concern	0	0	2	0
	Phyllomedusa bicolor		Least concern	0	11	0	0
	Phyllomedusa hypochrondialis		Least concern	0	0	0	1
Leptodactylidae	*Leptodactylus bolivianus*		Least concern	0	2	0	0
	Leptodactylus hyeri	X	Least concern	1	0	0	0
	Leptodactylus cf. *hylaedactylus*		Least concern	9	6	8	0
	Leptodactylus knudseni		Least concern	1	2	1	0
	Leptodactylus myersi		Least concern	0	13	0	0
	Leptodactylus mystaceus		Least concern	9	6	14	9
	Leptodactylus pentadactylus		Least concern	2	2	1	0
	Leptodactylus petersii		Least concern	0	3	1	0
	Leptodactylus rhodomystax		Least concern	0	0	9	1
Microhylidae	*Chiasmocleis shudikarensis*	X	Least concern	0	0	2	0
	Hamptophryne boliviana		Least concern	3	2	2	5
Pipidae	*Pipa aspera*		Least concern	0	0	1	0

table continued on next page

CLASS/(Sub)Order/Family	Species	Indicator of pristine forest	IUCN Red List Status	Kutari	Sipaliwini	Werehpai	Iwana Samu
Strabomantidae	*Pristimantis chiastonotus*	X	Least concern	0	2	1	0
	Pristimantis marmoratus	X	Least concern	0	0	2	0
	Pristimantis zeuctotylus		Least concern	1	8	0	1
GYMNOPHIONA							
Caeciliidae	*Microcaecilia taylori*	X	Least concern	1	0	0	0
TOTAL				**23**	**27**	**26**	**12**

REPTILIA

SERPENTES

Typhlopidae	*Typhlops reticulatus*		Least concern	1	0	0	0
Boidae	*Corallus enhydris*			0	0	1	0
	Eunectes murinus			0	0	1	0
Aniliidae	*Anillius scytale*			0	0	1	0
Colubridae	*Atractus flammigerus*			0	1	1	0
	Atractus torquatus			1	1	0	0
	Dipsas pavonina		Least concern	0	1	0	0
	Drymarchon corais			1	0	0	0
	Helicops angulatus			0	1	0	0
	Hydrops triangulatus			1	0	0	0
	Imantodes cenchoa			0	1	0	0
	Liophis typhlus			2	0	0	0
	Philodryas argenteus		Least concern	0	0	1	0
	Siphlophis cervinus			0	0	1	0
	Xenodon werneri			0	1	0	0
Viperidae	*Bothriopsis biliniatus*	X		0	0	1	0
	Bothrops atrox			0	1	0	0

AMPHISBAENIA

Amphisbaenidae	*Amphisbaena slevini*		Data deficient	1	0	0	0

SAURIA

Polychrotidae	*Anolis nitens*			2	4	3	0
	Anolis punctatus			0	0	1	0
Gekkonidae	*Coleodactylus amazonicus*			13	1	0	0
	Gonatodes annularis	X		0	0	2	0
	Gonatodes humeralis			0	0	2	0
	Thecadactylus rapicauda			0	0	1	1
Gymnophthalmidae	*Arthrosaura kocki*		Least concern	0	0	13	0
	Gymnopthalmus underwoodii		Least concern	0	0	1	0
	Leposoma guianense			8	2	17	0
	Neusticurus bicarinatus			1	1	0	0
Scincidae	*Mabuya nigropunctata*			1	0	3	0

table continued on next page

CLASS/(Sub)Order/Family	Species	Indicator of pristine forest	IUCN Red List Status	Kutari	Sipaliwini	Werehpai	Iwana Samu
Teiidae	*Kentropyx calcarata*			5	3	10	0
	Tupinambis nigropunctata			0	0	1	0
Tropiduridae	*Plica plica*			0	0	1	0
	Plica umbra			0	1	1	0
CHELONIA							
Chelidae	*Platemys platycephala*			0	0	1	0
Testudinidae	*Chelonoidis denticulata*		Vulnerable	1	0	0	0
CROCODILIA							
	Paleosuchus trigonatus		lower risk/ least concern (needs updating)	2	0	0	0
TOTAL				**14**	**13**	**21**	**1**

Chapter 10

Avifauna of the Kwamalasamutu region, Suriname

Brian J. O'Shea and Serano Ramcharan

SUMMARY

The RAP team recorded 327 species of birds: 294 species from the three RAP sites, 12 species observed in the area during the reconnaissance trip (3–8 May 2010) but not during the RAP survey, and 21 species observed only in the vicinity of Kwamalasamutu itself. The avifauna was typical of lowland forests of the Guiana Shield, and included many species endemic to the region. Our observations represent the first published records for Suriname of *Crypturellus brevirostris* (Rusty Tinamou), *Dromococcyx pavoninus* (Pavonine Cuckoo), *Xiphocolaptes promeropirhynchus* (Strong-billed Woodcreeper), and *Ramphotrigon megacephalum* (Large-headed Flatbill). The overall species list was highest for the Sipaliwini camp (250 species), followed by Werehpai (221 species) and Kutari (216 species). 153 species, or approximately 52% of those encountered at the three sites, were observed at all sites. The Kutari site had the most distinctive avifauna of the three sites. We estimate that a minimum of 350 bird species, or roughly half of the number of species known to occur in Suriname, may be found in the Kwamalasamutu area. Although no species listed on the IUCN Red List were encountered during the RAP survey, at least one (*Harpia harpyja*, Harpy Eagle, Near-Threatened) is known to occur in the area. Maintenance of large tracts of intact forest is recommended to preserve the avian diversity of the Kwamalasamutu region.

INTRODUCTION

Birds are excellent indicators for rapid biological assessments—they are primarily diurnal, they are generally easy to detect and identify, and the richness of bird communities tends to correlate positively with other measures of biodiversity. Birds are important food sources for other animals and people, and healthy populations of large-bodied frugivores and predators are indicative of a relatively intact, undisturbed ecosystem. Since many species are conspicuous when they are common, it is comparatively easy to assess their population status, even within the constraints of a rapid inventory.

In contrast to many other taxonomic groups, the avifauna of Suriname is well known (Ottema et al. 2009), though new records for the country continue to accumulate as more interior localities are inventoried (O'Shea 2005; Zyskowski et al. 2011). Most of the interior of Suriname is covered by unbroken tropical moist forest and is sparsely populated. Accordingly, the avifauna is diverse, and many sites support healthy populations of species that are of global conservation concern, such as large raptors, cracids, and parrots.

The Kwamalasamutu region encompasses the eastern portion of the upper Corantijn drainage in the southwest corner of Suriname. It is one of the most remote lowland regions of the Guiana Shield; much of the human population is concentrated in Kwamalasamutu itself, with human presence elsewhere limited to occasional hunting and fishing parties, or small groups of people traveling between communities along the major rivers. The region's vast forest matrix continues unbroken far into Brazil and Guyana, and is similarly isolated from the infrastructure of those countries. However, the planned construction of highways across northern Brazil and through the interior of Suriname poses a potential threat to the biodiversity of the Kwamalasamutu region. Illegal miners are a persistent presence throughout the interior regions of the Guianas, a situation that can be expected to worsen around Kwamalasamutu if roads allow easier land access to the region. As Suriname's infrastructure continues to develop, economic pressures on natural resources will increase. Therefore, the need to identify areas of exceptional biodiversity in Suriname is becoming ever more urgent.

We surveyed birds around three sites in the Kwamalasamutu area between 18 August and 8 September 2010. The purpose of the surveys was to obtain a baseline estimate of the avian species richness of the area, and to provide information on the population status of several bird species important to the Trio people. Our survey was preceded by an ornithological survey of the Werehpai area by Yale University in August 2006. Specimens from that expedition, representing many of the species listed in this report, are housed in the Peabody Museum of Natural History in New Haven, CT, USA.

STUDY SITES AND METHODS

We surveyed the avifauna at three localities in the Kwamalasamutu area between 19 August and 7 September 2010 (see Executive Summary (page 29) for site coordinates and Maps (page 13):

Site 1. Kutari River, 19–24 August.
Site 2. Sipaliwini River, 27 August–2 September.
Site 3. Werehpai, 3–7 September.

The habitat at all sites was a mosaic of tall terra firme and seasonally flooded forest, with the latter type most extensive at the Kutari River site. Within this mosaic were small patches of other habitat types, including so-called savanna forest, swamps dominated by *Euterpe oleracea* palms, xerophytic vegetation on granitic outcrops (inselbergs), and bamboo (*Guadua* sp.). Throughout the study period, we attempted to identify and survey as many different habitats as possible. The dates of the survey were chosen to fall within the long dry season, but the rainy season extended later than usual in 2010, and rain was frequent at the first site. Although local rainfall diminished substantially at the second and third sites, river levels remained high throughout the survey, indicating rain in the surrounding region. Birds were surveyed on foot for 1–2 hours before dawn, and during all morning hours of each day, primarily by walking along trails and identifying birds by sight and sound. We devoted most of our efforts to locating concentrations of birds or areas with good visibility, such as food sources (e.g., fruiting and flowering trees), mixed-species foraging flocks, or vantage points where large areas of canopy or sky could

be viewed. Birds were observed opportunistically at all other times of the day, generally in the vicinity of the camps.

Birds were documented using a Marantz PMD-661 digital recorder with a Sennheiser ME-62 omnidirectional microphone and Telinga parabolic reflector for individual birds, and a stereo microphone pair (Sennheiser MKH-20 and MKH-30) that was operated remotely for 2–3 hours at dawn on several mornings. Recordings are deposited at the Macaulay Library at the Cornell Lab of Ornithology in Ithaca, New York, USA.

RESULTS

Our list for the Kwamalasamutu region (Appendix) includes 332 species: 294 species were observed at the three camps, and 12 species were observed in the area during the reconnaissance trip (3–8 May 2010; wet season) but not during the RAP survey. We also include 21 species observed only in the vicinity of Kwamalasamutu itself; these species are probably restricted to the human-modified habitats around the village. Five additional species were not observed by us, but were documented with specimens and/or photographs by the Yale/Peabody expedition to Werehpai in 2006. We estimate that a minimum of 350 bird species, or roughly half of the number known to occur in Suriname, may be found in the Kwamalasamutu area.

The overall species list was highest for the Sipaliwini site (250 species), followed by Werehpai (221 species) and Kutari (216 species). 153 species, or approximately 52% of those encountered at the three sites, were observed at all sites. The Kutari site had the most distinctive assemblage of the three sites: although it had the fewest species, it had the most unique species (26) and shared fewer species with the Sipaliwini and Werehpai sites (180 and 163, respectively) than those sites shared with each other (203 species). Fifty species were observed at both the Sipaliwini and Werehpai sites but not at Kutari. The differences among sites were due in part to unequal distribution of certain habitats (e.g., *Guadua* bamboo, inselberg vegetation, river habitats) and their associated bird species (see below), but we attribute most of the differences to general rarity and the vagaries of sampling. This impression is corroborated by the observation that the majority of species not encountered at all sites are either relatively rare (e.g., birds of prey) or are most likely to be seen around widely dispersed resources that we were able to locate at some camps but not others (e.g., large fruiting trees). We therefore suspect that although the number of unique species at the Kutari site is indicative of habitat differences between forests along the Kutari and Sipaliwini Rivers, the majority of bird species reported here should be expected to occur at any of the survey sites, given additional sampling effort.

The avifauna of the Kwamalasamutu region was typical of lowland forests of the Guiana Shield. Of the 52 families encountered, three families of suboscine passerines (Furnariidae, Thamnophilidae, and Tyrannidae) accounted for over

30% of species observed. Due to the relative scarcity of fruiting and flowering trees during the survey, diversity of hummingbirds (Trochilidae) and tanagers (Thraupidae) was lower than expected. Although species composition was broadly similar among the three sites (more than half were observed at all sites), relative abundances of many species varied substantially among the camps. In particular, species that occur primarily or only in seasonally flooded forests were more common at the Kutari site, where this habitat type was most extensive. For other species, variation in abundance among sites may have been more apparent than real; for example, changes in singing behavior associated with the onset of the dry season may have made certain species seem more or less common as the RAP survey progressed. However, we suspect that most differences among sites could be attributed to variation in the distributions of microhabitats favored by particular species. Since many of these microhabitats are not stable over time in any particular place (e.g. treefall gaps), we do not consider our perceptions of variation in abundance to have any significant import for regional conservation.

We observed 15 species of parrots (Psittacidae). No species seemed especially common, and macaws (*Ara* spp.) were particularly scarce. Although larger species of parrots are hunted on an opportunistic basis, we could not attribute their low abundance at the time of the survey to hunting pressure. Parrots track their preferred food sources, and their abundance at a single site can vary dramatically over the course of a year—for example, two species were observed daily on the May reconnaissance trip but not at all during the RAP survey (see Appendix). The relative scarcity of parrots was likely an effect of limited food availability in the region at the time of our survey.

Guans (*Penelope* spp.) and especially Black Curassow (*Crax alector*), arguably the most important birds in the Trio diet, were less common in the Kwamalasamutu region than we have found them in other areas with little hunting pressure. Although they were observed at all of the sites, our records were often limited to second-hand reports and images from the camera traps.

NOTEWORTHY OBSERVATIONS

Four species represent new distributional records for Suriname: *Crypturellus brevirostris* (Rusty Tinamou), observed on two occasions at the Kutari site; *Dromococcyx pavoninus* (Pavonine Cuckoo), recorded remotely from the Sipaliwini site and observed daily in secondary growth around the Werehpai camp; *Xiphocolaptes promeropirhynchus* (Strong-billed Woodcreeper), recorded remotely at the Sipaliwini site; and *Ramphotrigon megacephalum* (Large-headed Flatbill), found in *Guadua* bamboo at both the Kutari and Werehpai sites. All four species are known from the Upper Essequibo region of extreme southern Guyana (Robbins et al. 2007; O'Shea 2008) but apparently do not occur farther north in the country (with the exception of *X.*

promeropirhynchus, which occurs in the tepui highlands). Although *C. brevirostris* and *D. pavoninus* are known to occur in adjacent northern Brazil, *R. megacephalum* is not; the Guyana record (O'Shea 2008) was the first for any of the Guianas and represented a 900-km range extension to the east (see Hilty 2003). Our observations further extend the range of this species, and we suspect it occurs in patches of *Guadua* bamboo elsewhere in the region.

NOTES ON SELECTED SPECIES

Harpy Eagle (*Harpia harpyja*; IUCN Near-Threatened): Although we did not observe Harpy Eagles during the RAP survey, the species is well known to the inhabitants of Kwamalasamutu, and the region undoubtedly supports a stable population. During the reconnaissance trip in May 2010, we were shown an abandoned nest site; this nest was active in August 2006, when it was photographed by the Yale/ Peabody expedition. In the Kwamalasamutu region and elsewhere in the Neotropics, Harpies are occasionally shot for food and other uses. As this species can be an excellent focal point for tourism, we recommend that they receive formal protection by the Trio.

Ciccaba virgata (Mottled Owl). We have included this species on the basis of a specimen collected at Werehpai by the Peabody Museum of Natural History in 2006, and by our own detections of unseen birds calling at both the Sipaliwini and Werehpai sites.

Asio stygius (Stygian Owl). This species is rare throughout its range, and is known from Suriname on the basis of several undocumented observations. We heard one individual near our camp at the Sipaliwini site, and we suspect that the species is a low-density resident throughout the forested interior of Suriname.

Nyctibius aethereus (Long-tailed Potoo) and *N. leucopterus* (White-winged Potoo). We heard both of these species (and recorded the latter) at the Kutari site. These potoos are rare and infrequently reported; in Suriname, the Kutari site is the third known locality for *N. aethereus* and the second for *N. leucopterus* (Ottema et al. 2009).

Deconychura longicauda (Long-tailed Woodcreeper). This species is rare in Suriname and appears to be absent from large areas of the country. We recorded a very vocal individual at the Kutari site.

Thamnophilus punctatus (Northern Slaty-Antshrike). This species was observed only on the inselberg at the Sipaliwini site, and in the Kwamalasamutu region it is probably restricted to the xerophytic vegetation typical of such rock outcrops.

Terenura cf. *spodioptila* ("Ash-winged" Antwren). We have provisionally included *T. spodioptila* on the list for the Kwamalasamutu region, as it is assumed to be the most widespread member of the genus *Terenura* in the Guiana Shield, most often observed in association with mixed-species foraging flocks in the canopy of tall forest, and usually detected by their distinctive songs; they are quite difficult to see well. Vocalizations of *Terenura* antwrens heard during the RAP survey, particularly from the Kutari site, could not be attributed with confidence to either *T. spodioptila* or *T. callinota*, which has been recorded from several sites in the interior of Suriname (usually on bauxite plateaus) and from the Acari Mountains of southern Guyana (O'Shea 2008; Zyskowski et al. 2011). More study of this genus in the Guiana Shield is needed.

Icterus cayanensis (Epaulet Oriole). This species is included on the basis of a remote recording of a singing bird at the Sipaliwini site.

CONSERVATION RECOMMENDATIONS

The Kwamalasamutu region is situated within a vast, intact block of tropical forest that faces no immediate threats. All of the species encountered on this survey also occur in the surrounding region, and the global populations of most are not threatened. However, some species, notably large-bodied predators and frugivores that require large areas of intact habitat for long-term population viability, probably maintain healthier populations here than elsewhere in their ranges. Care should be taken to preserve ecosystem integrity on the largest possible scale to forestall declines in their populations. To this end, the following guidelines should be adopted by the community of Kwamalasamutu:

- Aggressively exclude small-scale gold miners from Trio lands.
- Avoid trapping birds, particularly parrots, for export to coastal markets.
- Develop and implement a rotation system to distribute the effects of subsistence hunting over as large an area as possible; or, alternatively, designate more protected areas and enforce hunting bans.
- Increase production and consumption of domestic fowl as an alternative to bush meat.
- Enhance existing facilities to attract tourists to the area.

REFERENCES

Hilty, S.L. 2003. Birds of Venezuela, 2nd Edition. Princeton, NJ: Princeton University Press.

O'Shea, B.J. 2005. Notes on birds of the Sipaliwini savanna and other localities in southern Suriname, with six new species for the country. *Ornitología Neotropical* 16:361–370.

O'Shea, B.J. 2008. Birds of the Konashen COCA, southern Guyana. pp. 63–68 *in*: Alonso, L.E., J. McCullough, P. Naskrecki, E. Alexander, and H.E. Wright (eds.). A rapid biological assessment of the Konashen Community Owned Conservation Area, Southern Guyana. RAP Bulletin of Biological Assessment 51. Arlington, VA: Conservation International.

Ottema, O.H., J.H.J.M. Ribot, and A.L. Spaans. 2009. Annotated Checklist of the Birds of Suriname. Paramaribo: WWF Guianas.

Robbins, M.B., M.J. Braun, C.M. Milensky, B.K. Schmidt, W. Prince, N.H. Rice, D.W. Finch, and B.J. O'Shea. 2007. Avifauna of the upper Essequibo River and Acary Mountains, southern Guyana. *Ornitologia Neotropical* 18: 339–368.

Zyskowski, K., J. Mittermeier, O. Ottema, M. Rakovic, B.J. O'Shea, J.E. Lai, S.B. Hochgraf, J. de León, and K. Au. 2011. Avifauna of the easternmost tepui, Tafelberg in central Suriname. *Bulletin of the Peabody Museum of Natural History* 52:153–180.

Appendix. List of birds recorded from the Kwamalasamutu region, Suriname. Taxonomy, nomenclature, and linear sequence follow the current version of the American Ornithologists' Union South American Checklist (www.museum.lsu.edu/~Remsen/SACCBaseline.html). "R" indicates species observed only during the reconnaissance trip, 3–8 May 2010. "KW" indicates species observed only in or near the village of Kwamalasamutu. "Yale" denotes five species recorded from the Werehpai area in 2006 that were not seen by us; see Birds chapter for details.

Common name	Scientific name	Kutari	Sipaliwini	Werehpai	R	KW	Yale
TINAMIDAE							
Great Tinamou	*Tinamus major*	X	X	X			
Cinereous Tinamou	*Crypturellus cinereus*	X	X	X			
Little Tinamou	*Crypturellus soui*			X			
Variegated Tinamou	*Crypturellus variegatus*	X	X	X			
Rusty Tinamou	*Crypturellus brevirostris*	X					
CRACIDAE							
Guan sp.	*Penelope jacquacu/marail* sp.	X	X	X			
Blue-throated Piping-Guan	*Pipile cumanensis*		X	X			
Variable Chachalaca	*Ortalis motmot*	X	X	X			
Black Curassow	*Crax alector*	X	X	X			
ODONTOPHORIDAE							
Marbled Wood-Quail	*Odontophorus gujanensis*	X	X				
PHALACROCORACIDAE							
Neotropic Cormorant	*Phalacrocorax brasilianus*				X		
ANHINGIDAE							
Anhinga	*Anhinga anhinga*				X		
ARDEIDAE							
Rufescent Tiger-Heron	*Tigrisoma lineatum*	X	X	X			
Zigzag Heron	*Zebrilus undulatus*				X		
Striated Heron	*Butorides striata*	X	X	X			
Cattle Egret	*Bubulcus ibis*				X		
White-necked Heron	*Ardea cocoi*	X					
Capped Heron	*Pilherodius pileatus*		X	X			
THRESKIORNITHIDAE							
Green Ibis	*Mesembrinibis cayennensis*	X		X			
CATHARTIDAE							
Greater Yellow-headed Vulture	*Cathartes melambrotus*	X	X	X			
Black Vulture	*Coragyps atratus*		X	X			
King Vulture	*Sarcoramphus papa*	X	X	X			
PANDIONIDAE							
Osprey	*Pandion haliaetus*		X				
ACCIPITRIDAE							
Hook-billed Kite	*Chondrohierax uncinatus*				X		
Swallow-tailed Kite	*Elanoides forficatus*				X		
Plumbeous Kite	*Ictinia plumbea*		X				
White Hawk	*Pseudastur albicollis*	X	X				
Black-faced Hawk	*Leucopternis melanops*						X
Great Black Hawk	*Buteogallus urubitinga*		X				
Gray Hawk	*Buteo nitidus*		X	X			
Short-tailed Hawk	*Buteo brachyurus*					X	

table continued on next page

Common name	Scientific name	Kutari	Sipaliwini	Werehpai	R	KW	Yale
Harpy Eagle	*Harpia harpyja*						X
Black Hawk-Eagle	*Spizaetus tyrannus*		X	X			
Black-and-white Hawk-Eagle	*Spizaetus melanoleucus*		X				
Ornate Hawk-Eagle	*Spizaetus ornatus*		X				
FALCONIDAE							
Barred Forest-Falcon	*Micrastur ruficollis*	X	X				
Lined Forest-Falcon	*Micrastur gilvicollis*	X	X	X			
Slaty-backed Forest-Falcon	*Micrastur mirandollei*	X	X	X			
Collared Forest-Falcon	*Micrastur semitorquatus*	X	X	X			
Red-throated Caracara	*Ibycter americanus*	X	X	X			
Black Caracara	*Daptrius ater*		X	X			
Bat Falcon	*Falco rufigularis*	X	X				
PSOPHIIDAE							
Gray-winged Trumpeter	*Psophia crepitans*	X	X				
RALLIDAE							
Russet-crowned Crake	*Anurolimnas viridis*					X	
EURYPYGIDAE							
Sunbittern	*Eurypyga helias*		X	X			
CHARADRIIDAE							
American Golden-Plover	*Pluvialis dominica*					X	
SCOLOPACIDAE							
Spotted Sandpiper	*Actitis macularia*		X	X			
Solitary Sandpiper	*Tringa solitaria*		X	X			
COLUMBIDAE							
Common Ground-Dove	*Columbina passerina*					X	
Blue Ground-Dove	*Claravis pretiosa*					X	
Plumbeous Pigeon	*Patagioenas plumbea*	X	X	X			
Ruddy Pigeon	*Patagioenas subvinacea*		X				
White-tipped Dove	*Leptotila verreauxi*					X	
Gray-fronted Dove	*Leptotila rufaxilla*	X	X	X			
Ruddy Quail-Dove	*Geotrygon montana*				X		
PSITTACIDAE							
Blue-and-yellow Macaw	*Ara ararauna*	X	X				
Scarlet Macaw	*Ara macao*	X	X	X			
Chestnut-fronted Macaw	*Ara severus*		X	X			
Red-bellied Macaw	*Orthopsittaca manilata*					X	
White-eyed Parakeet	*Aratinga leucophthalma*		X	X			
Painted Parakeet	*Pyrrhura picta*	X	X	X			
Golden-winged Parakeet	*Brotogeris chrysoptera*	X	X	X			
Lilac-tailed Parrotlet	*Touit batavicus*				X		
Black-headed Parrot	*Pionites melanocephalus*	X	X				
Red-fan Parrot	*Deroptyus accipitrinus*	X	X	X			
Caica Parrot	*Pyrilia caica*	X	X				
Blue-headed Parrot	*Pionus menstruus*				X		

table continued on next page

Common name	Scientific name	Kutari	Sipaliwini	Werehpai	R	KW	Yale
Dusky Parrot	*Pionus fuscus*	X	X	X			
Orange-winged Parrot	*Amazona amazonica*	X	X				
Mealy Parrot	*Amazona farinosa*	X	X	X			
CUCULIDAE							
Little Cuckoo	*Coccycua minuta*	X					
Squirrel Cuckoo	*Piaya cayana*	X	X	X			
Black-bellied Cuckoo	*Piaya melanogaster*			X			
Cuckoo sp.	*Coccyzus cf. euleri*	X					
Smooth-billed Ani	*Crotophaga ani*					X	
Pavonine Cuckoo	*Dromococcyx pavoninus*		X	X			
STRIGIDAE							
Tawny-bellied Screech-Owl	*Megascops watsonii*	X	X	X			
Crested Owl	*Lophostrix cristata*	X	X	X			
Spectacled Owl	*Pulsatrix perspicillata*	X	X	X			
Mottled Owl	*Ciccaba virgata*		X	X			
Amazonian Pygmy-Owl	*Glaucidium hardyi*	X	X	X			
Stygian Owl	*Asio stygius*		X				
NYCTIBIIDAE							
Great Potoo	*Nyctibius grandis*		X	X			
Long-tailed Potoo	*Nyctibius aethereus*	X					
Common Potoo	*Nyctibius griseus*		X	X			
White-winged Potoo	*Nyctibius leucopterus*	X					
CAPRIMULGIDAE							
Short-tailed Nighthawk	*Lurocalis semitorquatus*	X	X				
Common Pauraque	*Nyctidromus albicollis*	X	X	X			
Blackish Nightjar	*Caprimulgus nigrescens*		X				
Ladder-tailed Nightjar	*Hydropsalis climacocerca*		X				
APODIDAE							
Band-rumped Swift	*Chaetura spinicaudus*	X	X	X			
Chapman's Swift	*Chaetura chapmani*		X	X			
Swift sp.	*Chaetura cf. meridionalis*		X				
TROCHILIDAE							
Crimson Topaz	*Topaza pella*						X
White-necked Jacobin	*Florisuga mellivora*	X	X	X			
Rufous-breasted Hermit	*Glaucis hirsutus*					X	
Pale-tailed Barbthroat	*Threnetes leucurus*						X
Reddish Hermit	*Phaethornis ruber*	X	X	X			
Straight-billed Hermit	*Phaethornis bourcieri*	X	X	X			
Long-tailed Hermit	*Phaethornis superciliosus*	X	X	X			
Black-eared Fairy	*Heliothryx auritus*		X				
Gray-breasted Sabrewing	*Campylopterus largipennis*		X				
Fork-tailed Woodnymph	*Thalurania furcata*	X	X	X			
Hummingbird sp.	*Amazilia cf. leucogaster*		X				
Rufous-throated Sapphire	*Hylocharis sapphirina*		X				
White-chinned Sapphire	*Hylocharis cyanus*		X	X			

table continued on next page

Common name	Scientific name	Kutari	Sipaliwini	Werehpai	R	KW	Yale
TROGONIDAE							
Black-tailed Trogon	*Trogon melanurus*	X	X	X			
Green-backed Trogon	*Trogon viridis*	X	X	X			
Guianan Trogon	*Trogon violaceus*	X	X	X			
Black-throated Trogon	*Trogon rufus*	X	X	X			
Collared Trogon	*Trogon collaris*	X	X	X			
ALCEDINIDAE							
Ringed Kingfisher	*Megaceryle torquata*	X	X	X			
Amazon Kingfisher	*Chloroceryle amazona*	X	X	X			
Green Kingfisher	*Chloroceryle americana*	X	X	X			
Green-and-rufous Kingfisher	*Chloroceryle inda*	X	X				
American Pygmy Kingfisher	*Chloroceryle aenea*	X	X				
MOMOTIDAE							
Amazonian Motmot	*Momotus momota*	X	X	X			
GALBULIDAE							
Brown Jacamar	*Brachygalba lugubris*	X	X	X			
Yellow-billed Jacamar	*Galbula albirostris*	X	X				
Green-tailed Jacamar	*Galbula galbula*		X				
Paradise Jacamar	*Galbula dea*	X	X	X			
Great Jacamar	*Jacamerops aureus*	X	X	X			
BUCCONIDAE							
Guianan Puffbird	*Notharchus macrorhynchos*	X	X				
Pied Puffbird	*Notharchus tectus*				X		
Collared Puffbird	*Bucco capensis*	X	X	X			
White-chested Puffbird	*Malacoptila fusca*	X	X				
Rusty-breasted Nunlet	*Nonnula rubecula*		X	X			
Black Nunbird	*Monasa atra*	X	X	X			
Swallow-winged Puffbird	*Chelidoptera tenebrosa*	X	X	X			
CAPITONIDAE							
Black-spotted Barbet	*Capito niger*	X	X	X			
RAMPHASTIDAE							
White-throated Toucan	*Ramphastos tucanus*	X	X	X			
Channel-billed Toucan	*Ramphastos vitellinus*	X	X	X			
Guianan Toucanet	*Selenidera culik*	X	X	X			
Green Aracari	*Pteroglossus viridis*	X	X	X			
Black-necked Aracari	*Pteroglossus aracari*	X	X	X			
PICIDAE							
Golden-spangled Piculet	*Picumnus exilis*	X		X			
Golden-collared Woodpecker	*Veniliornis cassini*	X	X	X			
Yellow-throated Woodpecker	*Piculus flavigula*	X	X	X			
Waved Woodpecker	*Celeus undatus*	X	X	X			
Chestnut Woodpecker	*Celeus elegans*	X					
Cream-coloured Woodpecker	*Celeus flavus*		X				
Ringed Woodpecker	*Celeus torquatus*		X				

table continued on next page

Common name	Scientific name	Kutari	Sipaliwini	Werehpai	R	KW	Yale
Lineated Woodpecker	Dryocopus lineatus		X	X			
Red-necked Woodpecker	Campephilus rubricollis	X	X	X			
Crimson-crested Woodpecker	Campephilus melanoleucos	X	X	X			
THAMNOPHILIDAE							
Fasciated Antshrike	Cymbilaimus lineatus	X	X	X			
Black-throated Antshrike	Frederickena viridis	X					
Great Antshrike	Taraba major		X	X			
Black-crested Antshrike	Sakesphorus canadensis		X	X			
Mouse-colored Antshrike	Thamnophilus murinus	X	X	X			
Northern Slaty-Antshrike	Thamnophilus punctatus		X				
Band-tailed Antshrike	Thamnophilus melanothorax		X	X			
Amazonian Antshrike	Thamnophilus amazonicus	X	X	X			
Dusky-throated Antshrike	Thamnomanes ardesiacus	X	X	X			
Cinereous Antshrike	Thamnomanes caesius	X	X	X			
Spot-winged Antshrike	Pygiptila stellaris				X		
Brown-bellied Antwren	Epinecrophylla gutturalis	X	X	X			
Pygmy Antwren	Myrmotherula brachyura	X	X	X			
Guianan Streaked-Antwren	Myrmotherula surinamensis	X	X	X			
Rufous-bellied Antwren	Myrmotherula guttata	X	X	X			
White-flanked Antwren	Myrmotherula axillaris	X	X	X			
Long-winged Antwren	Myrmotherula longipennis	X	X	X			
Gray Antwren	Myrmotherula menetriesii	X	X	X			
Spot-tailed Antwren	Herpsilochmus sticturus	X	X	X			
Todd's Antwren	Herpsilochmus stictocephalus	X	X	X			
Dot-winged Antwren	Microrhopias quixensis	X	X	X			
Guianan Warbling-Antbird	Hypocnemis cantator	X	X	X			
Ash-winged Antwren	Terenura spodioptila	X	X				
Gray Antbird	Cercomacra cinerascens	X	X	X			
Dusky Antbird	Cercomacra tyrannina	X	X	X			
White-browed Antbird	Myrmoborus leucophrys	X	X	X			
Black-chinned Antbird	Hypocnemoides melanopogon	X	X				
Silvered Antbird	Sclateria naevia	X					
Black-headed Antbird	Percnostola rufifrons	X	X	X			
Spot-winged Antbird	Schistocichla leucostigma	X	X				
Ferruginous-backed Antbird	Myrmeciza ferruginea	X	X	X			
Wing-banded Antbird	Myrmornis torquata	X	X				
White-plumed Antbird	Pithys albifrons	X	X	X			
Rufous-throated Antbird	Gymnopithys rufigula		X	X			
Spot-backed Antbird	Hylophylax naevius	X	X	X			
Scale-backed Antbird	Willisornis poecilinotus	X	X	X			
CONOPOPHAGIDAE							
Chestnut-belted Gnateater	Conopophaga aurita	X	X	X			
GRALLARIIDAE							
Variegated Antpitta	Grallaria varia		X				

table continued on next page

Common name	Scientific name	Kutari	Sipaliwini	Werehpai	R	KW	Yale
Spotted Antpitta	*Hylopezus macularius*	X	X	X			
Thrush-like Antpitta	*Myrmothera campanisona*	X	X	X			
FORMICARIIDAE							
Rufous-capped Antthrush	*Formicarius colma*	X	X	X			
Black-faced Antthrush	*Formicarius analis*	X	X	X			
FURNARIIDAE							
Short-billed Leaftosser	*Sclerurus rufigularis*	X					
McConnell's Spinetail	*Synallaxis macconnelli*						X
Rufous-rumped Foliage-gleaner	*Philydor erythrocercum*	X	X				
Cinnamon-rumped Foliage-gleaner	*Philydor pyrrhodes*			X			
Buff-throated Foliage-gleaner	*Automolus ochrolaemus*	X	X	X			
Olive-backed Foliage-gleaner	*Automolus infuscatus*	X	X	X			
Chestnut-crowned Foliage-gleaner	*Automolus rufipileatus*	X	X	X			
Rufous-tailed Xenops	*Microxenops milleri*	X					
Plain Xenops	*Xenops minutus*	X	X	X			
Plain-brown Woodcreeper	*Dendrocincla fuliginosa*	X	X	X			
Long-tailed Woodcreeper	*Deconychura longicauda*	X					
Wedge-billed Woodcreeper	*Glyphorynchus spirurus*	X		X			
Cinnamon-throated Woodcreeper	*Dendrexetastes rufigula*	X	X	X			
Red-billed Woodcreeper	*Hylexetastes perrotii*	X					
Strong-billed Woodcreeper	*Xiphocolaptes promeropirhynchus*		X				
Amazonian Barred-Woodcreeper	*Dendrocolaptes certhia*	X	X	X			
Black-banded Woodcreeper	*Dendrocolaptes picumnus*	X	X	X			
Striped Woodcreeper	*Xiphorhynchus obsoletus*	X					
Chestnut-rumped Woodcreeper	*Xiphorhynchus pardalotus*	X	X	X			
Buff-throated Woodcreeper	*Xiphorhynchus guttatus*	X	X	X			
Lineated Woodcreeper	*Lepidocolaptes albolineatus*	X					
Curve-billed Scythebill	*Campyloramphus procurvoides*		X	X			
TYRANNIDAE							
Sooty-headed Tyrannulet	*Phyllomyias griseiceps*	X		X			
Yellow-crowned Tyrannulet	*Tyrannulus elatus*	X	X	X			
Forest Elaenia	*Myiopagis gaimardii*	X	X	X			
Yellow-crowned Elaenia	*Myiopagis flavivertex*	X	X				
Elaenia sp.	*Elaenia cf. parvirostris*					X	
White-lored Tyrannulet	*Ornithion inerme*	X	X				
Southern Beardless-Tyrannulet	*Camptostoma obsoletum*	X	X	X			
Yellow Tyrannulet	*Campsiempis flaveola*					X	
Ringed Antpipit	*Corythopis torquatus*	X	X	X			
Guianan Tyrannulet	*Zimmerius acer*	X	X	X			
Olive-green Tyrannulet	*Phylloscartes virescens*	X					
Ochre-bellied Flycatcher	*Mionectes oleagineus*					X	
McConnell's Flycatcher	*Mionectes macconnelli*	X		X			
Short-tailed Pygmy-Tyrant	*Myiornis ecaudatus*	X	X	X			
Double-banded Pygmy-Tyrant	*Lophotriccus vitiosus*	X	X	X			
Helmeted Pygmy-Tyrant	*Lophotriccus galeatus*	X	X	X			

table continued on next page

Common name	Scientific name	Kutari	Sipaliwini	Werehpai	R	KW	Yale
Boat-billed Tody-Tyrant	*Hemitriccus josephinae*	X	X				
White-eyed Tody-Tyrant	*Hemitriccus zosterops*	X	X	X			
Common Tody-Flycatcher	*Todirostrum cinereum*					X	
Painted Tody-Flycatcher	*Todirostrum pictum*	X		X			
Yellow-margined Flycatcher	*Tolmomyias assimilis*	X	X	X			
Gray-crowned Flycatcher	*Tolmomyias poliocephalus*	X	X	X			
Cinnamon-crested Spadebill	*Platyrinchus saturatus*	X	X	X			
Golden-crowned Spadebill	*Platyrinchus coronatus*		X	X			
White-crested Spadebill	*Platyrinchus platyrhynchos*	X	X	X			
Royal Flycatcher	*Onychorhynchus coronatus*	X					
Bran-colored Flycatcher	*Myiophobus fasciatus*					X	
Sulphur-rumped Flycatcher	*Myiobius barbatus*	X	X				
Ruddy-tailed Flycatcher	*Terenotriccus erythrurus*	X	X				
Euler's Flycatcher	*Lathotriccus euleri*	X	X	X			
Drab Water Tyrant	*Ochthornis littoralis*		X	X			
Piratic Flycatcher	*Legatus leucophaius*		X	X			
Rusty-margined Flycatcher	*Myiozetetes cayanensis*		X	X			
Dusky-chested Flycatcher	*Myiozetetes luteiventris*	X	X	X			
Yellow-throated Flycatcher	*Conopias parvus*	X	X	X			
Sulphury Flycatcher	*Tyrannopsis sulphurea*					X	
Tropical Kingbird	*Tyrannus melancholicus*	X	X	X			
Grayish Mourner	*Rhytipterna simplex*	X	X	X			
Sirystes	*Sirystes sibilator*	X	X	X			
Dusky-capped Flycatcher	*Myiarchus tuberculifer*	X		X			
Short-crested Flycatcher	*Myiarchus ferox*	X	X	X			
Large-headed Flatbill	*Ramphotrigon megacephalum*	X		X			
Rufous-tailed Flatbill	*Ramphotrigon ruficauda*	X	X	X			
Cinnamon Attila	*Attila cinnamomeus*					X	
Bright-rumped Atilla	*Attila spadiceus*	X	X	X			
COTINGIDAE							
Guianan Red-Cotinga	*Phoenicircus carnifex*	X					
Guianan Cock-of-the-rock	*Rupicola rupicola*		X	X			
Purple-throated Fruitcrow	*Querula purpurata*	X	X	X			
Capuchinbird	*Perissocephalus tricolor*	X		X			
Spangled Cotinga	*Cotinga cayana*		X	X			
Screaming Piha	*Lipaugus vociferans*	X	X	X			
Pompadour Cotinga	*Xipholena punicea*		X				
Bare-necked Fruitcrow	*Gymnoderus foetidus*		X	X			
PIPRIDAE							
Tiny Tyrant-Manakin	*Tyranneutes virescens*	X	X	X			
White-throated Manakin	*Corapipo gutturalis*	X	X	X			
White-bearded Manakin	*Manacus manacus*		X	X			
White-crowned Manakin	*Pipra pipra*	X	X	X			
Golden-headed Manakin	*Pipra erythrocephala*	X	X	X			

table continued on next page

Common name	Scientific name	Kutari	Sipaliwini	Werehpai	R	KW	Yale
TITYRIDAE							
Black-tailed Tityra	*Tityra cayana*		X	X			
Thrush-like Schiffornis	*Schiffornis turdina*	X	X	X			
Cinereous Mourner	*Laniocera hypopyrra*	X	X				
Black-capped Becard	*Pachyramphus marginatus*	X	.	X			
Glossy-backed Becard	*Pachyramphus surinamus*	X			.		
INCERTAE SEDIS							
Wing-barred Piprites	*Piprites chloris*	X	X	X			
VIREONIDAE							
Rufous-browed Peppershrike	*Cyclarhis gujanensis*	X	X	X			
Slaty-capped Shrike-Vireo	*Vireolanius leucotis*	X	X	X			
Red-eyed Vireo	*Vireo olivaceus*	X	X	X			
Lemon-chested Greenlet	*Hylophilus thoracicus*	X	X	X			
Buff-cheeked Greenlet	*Hylophilus muscicapinus*	X	X	X			
Tawny-crowned Greenlet	*Hylophilus ochraceiceps*	X	X	X			
HIRUNDINIDAE							
Black-collared Swallow	*Pygochelidon melanoleuca*		X	X			
White-banded Swallow	*Atticora fasciata*	X	X	X			
Brown-chested Martin	*Progne tapera*		X	X			
Gray-breasted Martin	*Progne chalybea*		X	X			
White-winged Swallow	*Tachycineta albiventer*	X	X	X			
Bank Swallow	*Riparia riparia*					X	
Barn Swallow	*Hirundo rustica*		X	X			
Cliff Swallow	*Petrochelidon pyrrhonota*					X	
TROGLODYTIDAE							
Coraya Wren	*Pheugopedius coraya*	X	X	X			
Buff-breasted Wren	*Cantorchilus leucotis*	X	X	X			
Musician Wren	*Cyphorhinus arada*	X					
POLIOPTILIDAE							
Collared Gnatwren	*Microbates collaris*	X	X	X			
Long-billed Gnatwren	*Ramphocaenus melanurus*	X	X	X			
Tropical Gnatcatcher	*Polioptila plumbea*		X	X			
Guianan Gnatcatcher	*Polioptila guianensis*	X					
TURDIDAE							
Cocoa Thrush	*Turdus fumigatus*	X	X	X			
White-necked Thrush	*Turdus albicollis*		X	X			
THRAUPIDAE							
Red-capped Cardinal	*Paroaria gularis*		X	X			
Red-billed Pied Tanager	*Lamprospiza melanoleuca*	X	X				
Fulvous-crested Tanager	*Tachyphonus surinamus*	X	X	X			
White-shouldered Tanager	*Tachyphonus luctuosus*			X			
Fulvous Shrike-Tanager	*Lanio fulvus*	X	X	X			
Silver-beaked Tanager	*Ramphocelus carbo*	X	X	X			
Blue-gray Tanager	*Thraupis episcopus*					X	

table continued on next page

Common name	Scientific name	Kutari	Sipaliwini	Werehpai	R	KW	Yale
Palm Tanager	*Thraupis palmarum*		X	X			
Turquoise Tanager	*Tangara mexicana*		X	X			
Paradise Tanager	*Tangara chilensis*	X					
Opal-rumped Tanager	*Tangara velia*			X			
Swallow Tanager	*Tersina viridis*			X			
Black-faced Dacnis	*Dacnis lineata*		X	X			
Blue Dacnis	*Dacnis cayana*		X	X			
Purple Honeycreeper	*Cyanerpes caeruleus*				X		
Red-legged Honeycreeper	*Cyanerpes cyaneus*	X					
Green Honeycreeper	*Chlorophanes spiza*	X		X			
INCERTAE SEDIS							
Bananaquit	*Coereba flaveola*	X	X	X			
Slate-colored Grosbeak	*Saltator grossus*		X	X			
Buff-throated Saltator	*Saltator maximus*	X	X	X			
EMBERIZIDAE							
Blue-black Grassquit	*Volatinia jacarina*					X	
Pectoral Sparrow	*Arremon taciturnus*		X	X			
CARDINALIDAE							
Rose-breasted Chat	*Granatellus pelzelni*		X	X			
Yellow-green Grosbeak	*Caryothraustes canadensis*	X					
Blue-black Grosbeak	*Cyanocompsa cyanoides*		X	X			
PARULIDAE							
Tropical Parula	*Parula pitiayumi*	X					
Riverbank Warbler	*Phaeothlypis rivularis*	X					
ICTERIDAE							
Green Oropendola	*Psarocolius viridis*	X	X	X			
Crested Oropendola	*Psarocolius decumanus*	X	X	X			
Yellow-rumped Cacique	*Cacicus cela*	X	X	X			
Red-rumped Cacique	*Cacicus haemorrhous*		X	X			
Epaulet Oriole	*Icterus cayanensis*		X				
Giant Cowbird	*Molothrus oryzivorus*					X	
FRINGILLIDAE							
Euphonia sp.	*Euphonia cf. chlorotica*		X	X			
Violaceous Euphonia	*Euphonia violacea*	X	X	X			
Golden-sided Euphonia	*Euphonia cayennensis*	X	X	X			
Total species (332)		216	250	221			

Chapter 11

Rapid Assessment Program (RAP) survey of small mammals in the Kwamalasamutu region of Suriname

Burton K. Lim and Sahieda Joemratie

SUMMARY

In a Rapid Assessment Program (RAP) survey of the Kwamalasamutu region of southern Suriname, 38 species of small mammals were documented including 26 species of bats, 10 species of rats, and two species of opossums. The species diversity and relative abundance of rats at three sites around Kwamalasamutu were the highest recorded in 20 years of mammal surveys throughout Suriname and Guyana by the Royal Ontario Museum. Kutari was the most successful site for rats, indicating a healthy source of prey species for predators such as cats, owls, and snakes. In contrast, Werehpai was the most successful for bats but this was attributable to the well-established trails to the petroglyphs approximately 3.5 km from the river, which functioned as flyways that were more conducive for capture success compared to the other two sites where rudimentary trails were only recently cut. This indicates that bats are relatively tolerant to minor alternations to their habitat. Noteworthy records include two species endemic to the Guiana Shield, a water rat (*Neusticomys oyapocki*) and a brush-tailed rat (*Isothrix sinammariensis*), collected at Kutari that represent the first occurrences of these species in Suriname.

INTRODUCTION

Small mammals (bats, rodents, and opossums) comprise 80% of the mammalian species diversity in the Guianas (Lim et al. 2005). However, they are poorly known in comparison to the more charismatic and conspicuous larger species such as monkeys and cats. Approximately 200 species of mammals have been reported from Suriname. Small mammals are particularly important for conservation because many are fruit-eaters that disperse seeds necessary for natural forest succession, nectar-feeders that pollinate flowers, and insect-eaters that control natural populations through their foraging behavior and diet. High species diversity and relative abundance make small mammals an ideal group for rapid assessment program (RAP) surveys and long-term monitoring. This is particularly important for regions such as the Kwamalasamutu area that have not been thoroughly surveyed for biodiversity and conservation purposes.

STUDY SITES AND METHODS

We surveyed three sites in the Kwamalasamutu region: Kutari River (N 2.17538, W 56.78786), surveyed for six nights from 18–23 August; Sipaliwini (N 2.28979, W 56.60708), surveyed for five nights from 27–31 August; and Werehpai (N 2.36271, W 56.69860), surveyed for five nights from 2–6 September. Mist nets were also set at the petroglyph caves on the last night at the Werehpai site.

To survey non-volant small mammals during the RAP, we used Sherman live traps of two sizes: small (23 × 8 × 9 cm) and large (35 × 12 × 14 cm). Traps were set approximately five meters apart along transects on the ground near burrows, base of large trees, tree falls, along foraging runways, and rocky areas in upland forest and swamps for terrestrial animals. Some traps also were set on vines and low branches to sample arboreal species. Trapping effort per night varied among the sites with a maximum of 179 traps set at the Sipaliwini site.

Bats were captured with mesh mist nets 6 or 12 m in length that were set to a maximum height of 3 m in the forest understory. Pairs of short and long nets were set perpendicular approximately 100 meters apart along the transect across trails, over creeks, in swamps, near tree fall gaps, and by rocky outcrops where bats were typically flying. A maximum of 26 mist nets were set during the RAP and typically opened from approximately 1800 to 2400 h.

Small mammals not kept as representative samples of the species diversity in the Kwamalasamutu region were released unharmed at the point of capture. Individuals kept as voucher specimens were prepared as dried skins with carcasses temporarily preserved in ethanol for cleaning of the skulls and skeletons, or as whole animals fixed in 10% formalin with long-term storage in 70% ethanol. This will enable examination of both osteology and soft anatomy. Tissue samples of liver, heart, kidney, and spleen were frozen in the field with liquid nitrogen and for later storage in a −80°C ultra-cold freezer. Muscle samples were dabbed onto filter cards to stabilize DNA for sequencing in the international Barcode of Life project (www.barcodinglife.org) and also preserved in ethanol as a tissue backup precaution.

A reference collection of voucher specimens will be deposited at the University of Suriname's National Zoological Collection of Suriname, and the Royal Ontario Museum. Specimens will serve as documentation of the biodiversity of mammals in southern Suriname and will be available for study by the scientific community.

RESULTS

In total, preliminary field identifications indicated 38 species of small mammals represented by 375 individual captures, of which 251 were kept as voucher specimens (Appendix). More specifically, 26 species of bats were represented by 223 individuals (146 specimens kept as vouchers), 10 species of rats and mice were represented by 146 individuals (100 specimens), and two species of small opossums were represented by six individuals (five specimens). The overall species accumulation curve increased on every night except for one night (Fig. 1). This trend was primarily driven by the more species-rich bats because the non-volant small mammals had reached an asymptote by the fourth-last survey date.

In terms of individual sites, we documented 29 species of small mammals represented by 105 individuals at Kutari including 16 species of bats (52 individuals), 10 species of rats (52 individuals), and one species of opossum (one individual). At Sipaliwini we documented 22 species of small mammals represented by 84 individuals including 14 species of bats (47 individuals), five species of rats (36 individuals),

and one species of opossum (one individual). At Werehpai we documented 29 species of small mammals represented by 186 individuals including 23 species of bats (124 individuals), at least 5 species of rats (58 individuals), and one species of opossum (four individuals). Because the collecting permit limit of 100 rodent specimens was reached during the beginning of the Werehpai survey, individuals were released that could have potentially represented three additional species, so diversity at this site may be underestimated.

The species diversity and relative abundance of non-volant small mammals was substantially higher at Kutari than the other two sites (Figs. 2, 3). In contrast, the species diversity of bats was substantially higher at Werehpai (Fig. 4), as was the relative abundance (Fig. 5).

DISCUSSION

Although not many opossums were captured, five individuals of the short-tailed opossum (*Monodelphis brevicaudata*) were documented, which is the highest success rate in 20 years of similar surveys conducted in the Guianas by the Royal Ontario Museum. The short-tailed opossum is interesting in that it is active during the day (diurnal) searching for invertebrate prey such as insects and worms on the ground, whereas all other small mammals captured are active only at night (nocturnal).

The most common non-flying small mammal was the terrestrial rice rat (*Hylaeamys megacephalus*; page 23). A larger terrestrial rice rat (*Euryoryzomys macconnelli*; page 23) was the next most abundant. Rice rats are important seed predators in Neotropical rainforest. Spiny rats (*Proechimys* spp.; page 23) were also numerous and are one of the largest (up to 500 g) rats in South America. They are primary prey for many predators such as cats and snakes. Spiny mice (*Neacomys* spp.; page 23) were the next most abundant genus of non-volant small mammals.

For bats, the commonest species was the larger fruit-eating bat (*Artibeus planirostris*; page 23), which comprised almost one-third of all captures. This was also the most abundant species captured during the CI RAP surveys of the eastern Kanuku Mountains in Guyana (Lim and Norman, 2002), and Nassau and Lely Mountains in Suriname (Solari and Pinto, 2007). It is a fig (*Ficus*) specialist, however, the botanists found only a few fruiting fig trees during the survey, suggesting that either these bats rely on other fruits when figs are not masting or they are flying long distances from their day roost to fruiting fig trees. Other species in this genus (e.g., *A. jamaicensis* in Jalisco, Mexico) have been radio-tracked flying over 10 km in a night to feed at a fruiting fig tree (Morrison, 1978). The second most abundant species was the moustached bat (*Pteronotus parnellii*), which is an aerial insectivore. A common nectar-feeding bat that was caught at all 3 sites was *Lonchophylla thomasi* (page 23). The sword-nosed bat (*Lonchorhina inusitata*; page 23) was caught only at the Werehpai petroglyphs and may be dependent on caves or rock outcrops as roosting sites. It is an insect-feeding specialist.

The species diversity and relative abundance of rats and mice in the Kwamalasamutu area were the highest documented in 20 years of small mammal surveys throughout Suriname and Guyana by the Royal Ontario Museum. In particular, Kutari was the most successful site for rats and mice, indicating a healthy source of prey species for predators such as cats, owls, and snakes. Although the Kutari site is used occasionally as an overnight rest stop by Trio people traveling through the area, it is not inhabited, and is the least used of the three survey sites.

In contrast, Werehpai was the most successful for bats but this might be due in part to the well-established trail to the petroglyphs, which functioned as a flyway that was more conducive for capture success. The other two camps had lower bat diversity and abundance more typical of undisturbed forest, because transects were cut just before our arrival and were not functioning as flyways for bats. The Trio people recognize Werehpai as a protected area, but hunting still occurs on an irregular basis.

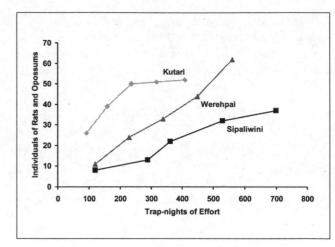

Figure 3. Individual accumulation curves for non-volant small mammals at the 3 survey sites in the Kwamalasamutu region of Suriname.

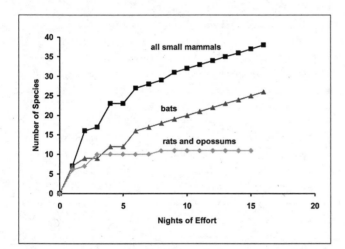

Figure 1. Species accumulation curves for small mammals in the Kwamalasamutu region of Suriname.

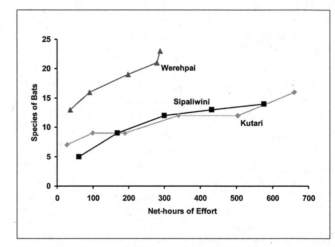

Figure 4. Species accumulation curves for bats at the 3 survey sites in the Kwamalasamutu region of Suriname.

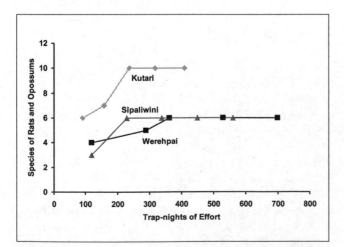

Figure 2. Species accumulation curves for non-volant small mammals at the 3 survey sites in the Kwamalasamutu region of Suriname.

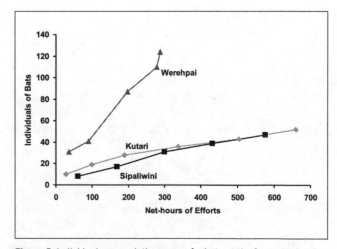

Figure 5. Individual accumulation curves for bats at the 3 survey sites in the Kwamalasamutu region of Suriname.

Sipaliwini had consistently low diversity and relative abundance of small mammals. Of the three surveyed areas, the Sipaliwini site is most often visited by hunters, primarily because of its proximity to Kwamalasamutu. In addition, trees or parts of trees, honey and other non-timber forest products are also harvested, so this area is more heavily used. In general, diversity and abundance of small mammals appeared to correlate negatively with the amount of human disturbance in the three areas sampled.

Interesting Species

A water rat (*Neusticomys oyapocki*) was collected at Kutari that represents the first documentation of this species in Suriname. The ears and eyes are reduced in size as an adaptation for aquatic behaviour. There are only 10 specimens of this species (Leite et al. 2007) and not much is known of its ecology or role in the ecosystem.

Another interesting species was a brushed-tailed rat (*Isothrix sinnamariensis*) that was found by the large mammal camera trapping team. It was discovered dead with wounds on the head and shoulders on a part of the trail they had just recently walked. Indications are that they had startled a raptor that had killed the rat, which was then dropped on the trail. This represents the first report from Suriname of a brush-tailed rat, an arboreal species endemic to the Guianas and known from fewer than 8 specimens (Lim et al. 2006; Patterson and Velazco 2008). With these faunal additions, there are currently 196 species of mammals documented from Suriname (Lim et al. 2005; Lim 2009).

CONSERVATION RECOMMENDATIONS

The Kutari site was furthest from the village of Kwamalasamutu and the most remote of the three camps, which may partially account for the high species diversity and relative abundance of rats and mice. This taxonomic group is primary prey for many top-level nocturnal predators such as cats, snakes, and owls. A healthy predator-prey relationship is a good indicator of the conservation status of forest habitat. Kutari would be a good candidate area for a nature reserve within the Kwamalasamutu region.

Sipaliwini had the lowest species diversity and relative abundance, but also the most homogeneous forest habitat. A variety of microhabitats such as swamp forest and rocky outcrops were present near the other sites, and usually are associated with more diverse and abundant small mammal faunas.

Werehpai had the highest bat diversity and abundance suggesting that this taxonomic group adapts well to minor habitat changes such as the establishment and maintenance of trails in the forest. Flyways act as convenient routes within the forest for greater access to food resources such as fruits, flowers, and insects. However, over-development (such as construction of permanent buildings) causes changes to the community ecology of bats and alters their impact on the environment in terms of forest composition associated with seed dispersal and pollination, as documented at the ecotourism resort of Blanche Marie Vallen in western Suriname (Lim 2009).

The primary conservation recommendations arising from the small mammal survey of the Kwamalasamutu region are: 1) designation of the Kutari area as a nature reserve because of the high species diversity and relative abundance of rats and mice that are necessary to sustain healthy populations of top-level predators; and 2) minimal development of the Werehpai petroglyph site to ensure continued ecosystem services of the bat fauna including seed dispersal, flower pollination, and insect control.

REFERENCES

Leite, R.N., M.N.F. da Silva, T.A. Gardner. 2007. New records of *Neusticomys oyapocki* (Rodentia, Sigmodontinae) from a human-dominated forest landscape in northeastern Brazilian Amazonia. Mastozoologia Neotropical 14: 257–261.

Lim, B.K. 2009. Environmental assessment at the Bakhuis bauxite concession: small-sized mammal diversity and abundance in the lowland humid forests of Suriname. The Open Biology Journal, 2:42–53.

Lim, B.K., M.D. Engstrom, and J. Ochoa G. 2005. Mammals. *In*: Checklist of the terrestrial vertebrates of the Guiana Shield (T. Hollowell and R. P. Reynolds, eds.). Bulletin of the Biological Society of Washington. 13: 77–92.

Lim, B.K., M.D. Engstrom, J.C. Patton, and J.W. Bickham. 2006. Systematic relationships of the Guianan brush-tailed rat (*Isothrix sinnamariensis*) and its first occurrence in Guyana. Mammalia, 70:120–125.

Lim, B.K., and Z. Norman. 2002. Rapid assessment of small mammals in the eastern Kanuku Mountains, Lower Kwitaro River area, Guyana. Pp. 51–58, in Montambault, (J.R. and O. Missa (eds.). A biodiversity assessment of the eastern Kanuku Mountains, Lower Kwitaro River, Guyana Conservation International, RAP Bulletin of Biological Assessment, 26. Conservation International, Arlington, VA, USA.

Morrison, D.W. 1978. Influence of habitat on the foraging distances of the fruit bat, *Artibeus jamaicensis*. Journal of Mammalogy, 59: 622–624.

Patterson, B.D., and P.M. Velazco. 2008. Phylogeny of the rodent genus *Isothrix* (Hystricognathi, Echimyidae) and its diversification in Amazonia and the eastern Andes. Journal of Mammalian Evolution, 15:181–201.

Solari, S., and M. Pinto. 2007. A rapid assessment of mammals of the Nassau and Lely plateaus, Eastern Suriname. Pp. 130–134, in Alonso, L.E. and J.H. Mol (eds.). A rapid biological assessment of the Lely and Nassau plateaus, Suriname (with additional information on the Brownsberg Plateau). RAP Bulletin of Biological Assessment 43. Conservation International, Arlington, VA, USA.

Appendix. List of mammals collected on the Kwamalasamutu RAP survey. Number of individuals collected (in parentheses indicates total individuals, including individuals that were released).

Species	Kutari	Sipaliwini	Werehpai	Individuals
Opossums:				
Marmosops parvidens	1			1
Monodelphis brevicaudata		1	3 (4)	4 (5)
Subtotal	**1**	**1**	**3 (4)**	**5 (6)**
Rats:				
Neusticomys oyapocki	1			1
Isothrix sinammariensis	1			1
Neacomys guianae	2			2
Neacomys paracou	1		1 (9)	2 (10)
Oecomys auyantepui	4	2	1 (2)	7 (8)
Oecomys rutilus	1			1
Euryoryzomys macconnelli	6	12	1 (10)	19 (28)
Hylaeamys megacephalus	28	14	7 (30)	49 (72)
Proechimys cuvieri	1	4	2 (7)	7 (12)
Proechimys guyannensis	7	4		11
Subtotal	**52**	**36**	**12 (58)**	**100 (146)**
Bats:				
Anoura geoffroyi		1		1
Artibeus bogotensis	1		1	2
Artibeus gnomus	2		1	3
Artibeus lituratus	1	3	3 (12)	7 (16)
Artibeus obscurus	3	3	3 (6)	9(12)
Artibeus planirostris	14 (16)	9	7 (45)	30 (70)
Carollia brevicauda	1		2	3
Carollia perspicillata	2	2	1 (2)	5 (6)
Desmodus rotundus	1		2	3
Lionycteris spurrelli		2	2 (6)	4 (8)
Lonchophylla thomasi	3	2	2(4)	7 (9)
Lonchorhinua inusitata			2 (8)	2 (8)
Lophostoma silvicolum	6	8	1	15
Micronycteris megalotis			1	1
Micronycteris minuta		1		1
Mimon crenulatum		1	2	3
Myotis riparius	1	1	1	3
Phyllostomus discolor			1	1
Phyllostomus elongatus	4	1	5 (8)	10 (13)
Phyllostomus hastatus			1	1
Platyrrhinus helleri	2		2	4
Pteronotus parnellii	6	11	3 (10)	20 (27)
Rhinophylla pumilio	1		3 (5)	4 (6)

table continued on next page

Species	Kutari	Sipaliwini	Werehpai	Individuals
Sturnira tildae			1	1
Trachops cirrhosus	2		2	4
Uroderma bilobatum		2		2
Subtotal	**50 (52)**	**47**	**49 (124)**	**146 (223)**
TOTAL	**103 (105)**	**84**	**64 (186)**	**251 (375)**

Chapter 12

A survey of the large mammal fauna of the Kwamalasamutu region, Suriname

*Krisna Gajapersad, Angelique Mackintosh,
Angelica Benitez, and Esteban Payán*

INTRODUCTION

Historically, humans have used animals for food and a variety of other uses (Leader-Williams et al. 1990; Milner-Gulland et al. 2001). Examples all over the world show the effects of over-hunting from humans, causing population declines and extinction (Diamond 1989). Overexploitation was almost certainly responsible for historical extinctions of some large mammals and birds (Turvey and Risley 2006). Large mammals are more sensitive to hunting due to their slow reproductive rates, long development and growth times, and large food and habitat requirements (Purvis et al. 2000; Cardillo et al. 2005). Today, roughly two million people depend on wild meat for food or trade (Fa et al. 2002; Milner-Gulland et al. 2003), yet the majority of hunting is unsustainable (Robinson and Bennett 2004; Silvius et al. 2005).

Subsistence hunting of terrestrial vertebrates is a widespread phenomenon in tropical forests (Robinson and Bennett 2000). In many parts of Latin America, cracid (Aves: Cracidae) populations are declining (Thiollay 2005). Subsistence hunting is an important cause of these declines (Thiollay 1989; Ayres et al. 1991; Silva and Strahl 1991; Strahl and Grajal 1991; Vickers 1991; Hill et al. 2003). The direct impacts of hunting on animal populations and the subsequent effects of exploitation on the ecosystem make attaining sustainable harvests an international conservation priority (Fa et al. 2003; Milner-Gulland et al. 2003; Bennett et al. 2007). Thus, the first step in making harvests more sustainable is to determine current levels of harvest (Milner-Gulland and Akcakaya 2001).

Mammals as a group provide the main protein source for indigenous peoples of Amazonia. Indigenous tribes have lived in Amazonia for tens of thousands of years (Redford 1992) and many, including the Trio of Suriname, still remain within the forest and hunt mammals actively. Abundances of large mammals have decreased in areas where they have been hunted (Peres 1990; Cullen et al. 2000; Hill et al. 2003). Unmanaged hunting is commonplace in the Amazon and tends to deplete game populations, often to levels so low that local extinctions are frequent (Redford 1992; Bodmer et al. 1994). Overhunting then becomes a double-edged threat: to the biodiversity of the tropics and to the people that depend on those harvests for food and income.

At the present time, little information is available on the occurrence, spatial variability in richness, and sensitivity to hunting and other disturbances of medium and large mammals in Suriname. The goal of this survey was to assess the diversity and abundance of medium- and large-bodied mammals in the Kwamalasamutu region.

METHODS AND STUDY SITES

We surveyed medium- and large-bodied mammals by means of three main methods: camera trapping, searching for scat and animal tracks, and making visual and aural observations. We also characterized hunting habits of the Trio through interviews with residents of Kwamalasamutu.

Camera traps were set 500 meters apart along hunting and game trails, some of which were cut shortly before the RAP survey. The camera traps operated day and night, photographing all ground-dwelling mammals and birds that walked in front of them. Camera traps were attached to trees approximately 30 cm above the forest floor.

At the Kutari site 25 camera traps were set up, divided over 4 trails, and run for a total of 181 camera trap days. At the Sipaliwini site 12 camera traps were set up, divided over 3 trails, and were active for a total of 104 camera trap days. At Werehpai, there was no trail cutting due to preexisting trails and 10 camera traps were set up along 2 trails and run for 304 camera trap days. Cameras were placed in different habitats at each of the study sites. At the Kutari site 15 camera traps were set up in terra firme, five in swamp, four in flooded forest and one in a dry creek bed. At the Sipaliwini site nine camera traps were set up in terra firme, two in swamp and one in a creek. At the Werehpai site, eight cameras were set up in terra firme and two near creeks. Elevations of camera trapping points were similar among the three sites, ranging between 213 and 278 meters.

Photographs from camera traps were identified to species, and independent photographs were used as single occurrences for analysis. Independent photographs were those from different species or individuals, or any photographs taken at least 30 minutes apart (O'Brien et al. 2003). Rarefied species accumulation curves and biodiversity indices were calculated with program EstimateS 8.2 (Colwell 2009). Occurrences from photographs were compared with the nonparametric richness estimator Chao 1 among camps, and Simpson's Biodiversity Index was also calculated per camp (Magurran 2004). Additionally, a relative abundance index was estimated per species for every 100 trap-nights (O'Brien et al. 2003).

Tracks and scat were also recorded when walking the trails to set up and pick up the camera traps. The tracks were identified with the help of local guides that accompanied the field excursions, and the tracks that could not be identified in the field were photographed and identified with the help of field guides. Visual and aural observations were important for the primates, because this group of animals is not captured by the camera traps, have diurnal habits and do not leave tracks on the forest floor. Interviews were conducted with hunters and elders from the area. We sought information on hunting habits, frequency, weapons, and the abundance of preferred and actual prey.

RESULTS

We detected 29 species of medium- and large-bodied mammals (Appendix). We recorded 22 mammal species from the Kutari site, including all eight primate species that occur in Suriname. At the Sipaliwini site we found 18 mammal species, including four primate species; at Werehpai we found 21 mammal species including five primate species.

The large caviomorph rodents, especially Paca (*Cuniculis paca*), Red-rumped Agouti (*Dasyprocta leporina*) and Red Acouchy (*Myoprocta acouchy*), were the most frequently photographed by the camera traps (Table 1); this group was assumed to include the most common medium- and large-bodied nonvolant mammals in the area. The Brazilian Tapir (*Tapirus terrestris*) was recorded by the camera traps at all three sites and was observed by several of the RAP scientists. A large number of tracks were found on the trails, indicating that the Brazilian Tapir is common in the area.

Of the six species of cats known to occur on the Guiana Shield, the Jaguar (*Panthera onca*), Puma (*Puma concolor*) and Ocelot (*Leopardus pardalis*) were found during the survey. Ocelot was the most frequently recorded cat species during this survey and is common in the area. The Jaguar and Puma were each recorded by the camera traps only once, both in the Werehpai area (Table 1). Tracks of Puma were also found at the Kutari site. It is very likely that the Jaguar also occurs in the Kutari and Sipaliwini area, but was only recorded in the Werehpai area because the trail system at Werehpai is used frequently by large cats. The Trio do not actively hunt cats, but they occasionally kill the large cats when they encounter them in the forest, because they are afraid of being attacked.

In all three camps both the Red-brocket and Grey-brocket Deer (*Mazama americana* and *M. gouazoubira*) were recorded by the camera traps and detected by tracks. Tracks of the Collared Peccary (*Pecari tajacu*) were found at all 3 camps, and this species was also recorded frequently by the camera traps. The White-lipped Peccary (*Tayassu pecari*) was only photographed once by the camera traps in the Werehpai area, and seems to be uncommon in the Kwamalasamutu region.

Three armadillo species were found during the RAP: Great Long-nosed Armadillo (*Dasypus kappleri*), Nine-banded Armadillo (*Dasypus novemcinctus*), and Giant Armadillo (*Priodontes maximus*). The Giant Anteater (*Myrmecophaga tridactyla*) was recorded by the camera traps only once at the Kutari site. Four species of ground-dwelling birds were recorded by the camera traps and observed during the RAP: Black Curassow (*Crax alector*), Grey-winged Trumpeter (*Psophia crepitans*), Variegated Tinamou (*Crypturellus variegatus*), and Great Tinamou (*Tinamus major*).

All surveys were incomplete. Species accumulation curves from photographs at the three sites show that more species could be expected to occur at the sites (Fig. 1). The Chao 1 diversity estimator confirms this, showing the expected number of species available for detection (Fig. 2). Chao 1 estimates that the survey of the Kutari site was close to completion, with less than 5% of expected species remaining to be detected. In contrast, the Sipaliwini and Werehpai site surveys appeared to be far from complete, with more than 27% of expected species at Sipaliwini and 53% at Werehpai remaining to be detected. The slope of the curve denotes the detection rate, which was highest at the Sipaliwini site (m = 0.9), followed by the Kutari site (m = 0.5) and finally Werehpai (m = 0.2; Fig. 2). This is congruent with Simpson's diversity index, for which Werehpai had the least even species abundance distribution (d = 3.4), whereas the Kutari and Sipaliwini sites had very similar and more even abundances of species (d = 8.7 and d = 9.5, respectively).

Table 1. Total independent photographs and relative abundance indices (RAI) for all vertebrate species detected by camera traps per RAP site.

	Kutari		Sipaliwini		Werehpai	
	Photos	RAI	Photos	RAI	Photos	RAI
Proechymis sp.	9	7.2	4	4.5	4	1.6
Neacomys sp.	3	2.4	1	1.1	1	0.4
Cuniculus paca	5	4	7	7.9	10	3.9
Dasyprocta leporina	6	4.8	6	6.7	1	0.4
Myoprocta acouchy	19	15.2	4	4.5	10	3.9
Dasypus kappleri	2	1.6	2	2.2	1	0.4
Dasypus novemcinctus	2	1.6	0	0.0	1	0.4
Priodontes maximus	0	0	1	1.1	2	0.8
Mazama americana	4	3.2	2	2.2	4	1.6
Mazama gouazoubiria	1	0.8	2	2.2	0	0.0
Metachirus nudicaudatus	3	2.4	0	0.0	2	0.8
Philander opposum	2	1.6	0	0.0	0	0.0
Didelphis marsupialis	3	2.4	1	1.1	0	0.0
Psophia crepitans	20	16	0	0.0	0	0.0
Pecari tajacu	4	3.2	2	2.2	0	0.0
Tapirus terrestris	2	1.6	0	0.0	2	0.8
Eira barbara	0	0	0	0.0	1	0.4
Nasua nasua	1	0.8	0	0.0	1	0.4
Leopardus pardalis	3	2.4	0	0.0	7	2.7
Panthera onca	0	0	0	0.0	1	0.4
Puma concolor	0	0	0	0.0	1	0.4
Total photos	**89**		**32**		**49**	

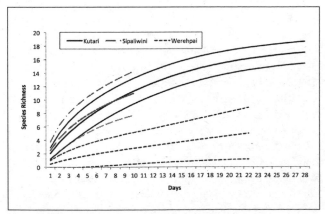

Figure 1. Observed species accumulation curves with confidence intervals (95%; upper and lower) from camera trap pictures.

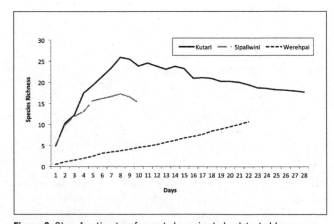

Figure 2. Chao 1 estimator of expected species to be detected by camera traps at each RAP survey site.

INTERVIEWS

The interview data provided an overview of the hunting areas, hunting techniques, frequency of hunting, and species hunted. All hunters interviewed were men from the village of Kwamalasamutu. Hunting techniques are generally learned at the age of 10–15 from the father or grandfather. Most of the interviewed men first learned to hunt with bow and arrow and later (at age 14–16) with a gun. All of the interviewed men hunt to supply their families with food, and sometimes the meat is sold on the market or given to other family members. They normally hunt once a week, and a hunting trip generally lasts one day and one night (24 hours). Hunting is done alone or with a family member or friend. Most of the hunters first go two hours by boat upstream or downstream from Kwamalasamutu, and then walk several hours into the forest to hunt. Black curassow (*Crax alector*) was the preferred species among interviewees. Other large ground-dwelling bird species, such as tinamous (*Tinamus major*, *Crypturellus* spp.) and Grey-winged Trumpeters (*Psophia crepitans*), were also favorites. Paca (*Cuniculus paca*), Collared Peccary (*Pecari tajacu*), and Red Acouchy (*Myoprocta acouchy*) were the preferred mammal species.

The most hunted mammal is the Guianan Red Howler Monkey (*Alouatta macconnelli*), because this species is easily spotted in trees along the river. Other frequently hunted species are White-lipped Peccary (*Tayassu pecari*), Red-rumped Agouti (*Dasyprocta leporina*), Red Acouchy (*Myoprocta acouchy*), Paca (*Cuniculus paca*), and Black Curassow (*Crax alector*). Large mammals that are less common, but hunted for food when encountered in the forest, are Brazilian Tapir (*Tapirus terrestris*), Red-brocket Deer (*Mazama americana*), Giant Armadillo (*Priodontes maximus*), and Giant Anteater (*Myrmecophaga tridactylus*). With the exception of Red-brocket Deer, all of these animals are listed on the IUCN Red List of Threatened Species.

Some of the interviewed hunters also mentioned that they use a traditional hunting calendar and hunt different species in different seasons. These hunters said that they do not hunt when the animals are "fat"; meaning that they do not kill animals when it is visible that they are in the gestation period.

CONSERVATION RECOMMENDATIONS

The number of mammal species found during this survey does not differ much from what was expected. Most of the large terrestrial mammal species expected to occur in the region were recorded by the camera traps. The difference in number of species per site suggests that hunting pressure in the different areas varies. The Kutari site was the richest in species, especially primates, suggesting limited hunting pressure. This site also had a high value of Simpson's Index from the camera trap data, indicating a high diversity and abundance of medium- and large-bodied terrestrial mammals,

as well as presence of some sensitive bird species such as Gray-winged Trumpeter (*Psophia crepitans*), and the highest relative abundance index for Brazilian Tapir of any of the three sites (Table 1). We attribute the richness of the Kutari mammal fauna to its isolation from Kwamalasamutu, relative to the Sipaliwini and Werehpai sites. Of the three sites, the Sipaliwini site had the highest value of Simpson's Index, but also the smallest number of species recorded by camera traps, tracks and observations, suggesting higher hunting pressure in the area. The value of Simpson's Index for this site was a result of the even abundance of several species of rodents, as well as deer (*Mazama* spp.), which tolerate disturbance quite well; only one photograph of a sensitive species (Giant Armadillo) was obtained at this site. This area is used as a hunting ground by the local people, and hunting trails were encountered during camera trap setup. During the RAP, several shots from hunters were heard near the Sipaliwini camp. The Werehpai site was within the indigenous protected area established by the local village authority in 2004. We found more species at Werehpai than at Sipaliwini, even though it is only ten kilometers from Kwamalasamutu. Werehpai had a low value of Simpson's Index, due to the high abundance of two species of rodents, but we did record both Brazilian Tapir and Jaguar at this site.

The results of this survey suggest that richness and evenness of the medium- and large-bodied mammal fauna both increase with distance from Kwamalasamutu. Nevertheless, the presence of species sensitive to hunting and disturbance, such as tapir, jaguar, curassows and large primates, suggests that hunting pressure is not pervasive. Hunting is probably limited by reduced river access to some areas in the dry season, and more generally by distance from Kwamalasamutu. The concentration of the Trio in Kwamalasamutu reduces hunting pressure on large vertebrates in the region as a whole. The extensive surrounding forest acts as a source to offset local population depletion due to hunting, up to a point. The current village is relatively large with an estimated 700–800 people who all depend on the surrounding forest for sustenance. This puts much pressure on the medium- and large-bodied mammals in the area surrounding the village. Our interview results indicate that the effort required to find desired prey (i.e. large vertebrates) is increasing, and hunters reported that they often have to travel far from Kwamalasamutu to hunt successfully. Prey depletion around the edges of hunting villages is an expected phenomenon. The size of the area affected by hunting can be expected to increase as human population density forces more frequent long-distance hunting expeditions. Therefore, uncontrolled hunting from Kwamalasamutu represents the most significant potential threat to the mammal species in the region. Declaring the Werehpai area as a protected area is a good initiative by the village authority to conserve the species on which they depend for food, but this is only a small area compared to the hunting areas. More monitoring is required to determine if the Werehpai protected area is sufficient to maintain populations of medium- and large-bodied mammals in the surroundings of Kwamalasamutu. Further, it is recommended that all hunters from Kwamalasamutu use hunting seasons, map zoning areas and set quotas for hunting on the different species to maximize long-term viability of game populations in the area. These hunting seasons should be developed together with the people from Kwamalasamutu, with traditional local knowledge augmenting a scientific approach. Zoning must be established to achieve population source (e.g. game reserves) and sink (i.e. hunting) areas for wild meat, ensuring a permanent supply for subsistence (Novaro 2004). Sale of meat should be restricted to the village. Hunting quotas should be based on sustainability measures (Wilkie et al. 1998; Robinson and Bennett 2004; Payan 2009) and harvest profiles. Small- and medium-bodied rodents and species with high reproductive rates should be favored, whereas hunting of large animals (e.g., tapir) and other less resilient species should be highly controlled (Bodmer 1995).

A more thorough study of the medium- and large-bodied mammal fauna of the Kwamalasamutu region is recommended, including a longer camera trapping study and a detailed sustainability evaluation of wild meat hunting.

REFERENCES

Alvard, M. (1995). Shotguns and sustainable hunting in the Neotropics. *Oryx* 29(1): 58–66.

Ayres, J., D. Lima, E. Martins, and J. Barreiros. (1991). On the track of the road: changes in subsistence hunting in a Brazilian Amazonian Village. Neotropical Wildlife Use and Conservation. J. G. Robinson and K. H. Redford, eds. Univeristy of Chicago Press.

Bennett, E. L., E. Blencowe, K. Brandon, D. Brown, R.W. Burn, G. Cowlishaw, G. Davies, H. Dublin, J.E. Fa, E.J. Milner-Gulland, J.G. Robinson, J.M. Rowcliffe, F.M. Underwood and D.S. Wilkie. (2007). Hunting for Consensus: Reconciling Bushmeat Harvest, Conservation, and Development Policy in West and Central Africa. *Conservation Biology* 21(3): 884–887.

Bodmer, R. (1995). Managing Amazonian wildlife: Biological correlates of game choice by detribalized hunters. *Ecological Applications* 5(4): 872–877.

Bodmer, R. E., T. Fang, L. Moya and R. Gill. (1994). Managing wildlife to conserve Amazonia forests: population biology and economic considerations of game hunting. *Biological Conservation* 67: 29–35.

Cardillo, M., G.M. Mace, K.E. Jones, J. Bielby, O.R. Bininda-Emonds, W. Sechrest, C.D. Orme, A. Purvis. (2005). Multiple Causes of High Extinction Risk in Large Mammal Species. *Science* 309(5738): 1239–1241.

Colwell, R. 2009. EstimateS: Statistical estimation of species richness and shared species from samples. Viceroy University.

Cullen, L., R.E. Bodmer, and C.V. Pa. (2000). Effects of hunting in habitat fragments of the Atlantic forests, Brazil. *Biological Conservation* 95(1): 49–56.

Diamond, J. (1989). The Present, Past and Future of Human-Caused Extinctions. *Philosophical Transactions of the Royal Society of London. Series B, Biological Sciences* (1934–1990) 325(1228): 469–477.

Dinata, Y., A. Nugroho, I.A. Haidir, and M. Linkie. (2008). Camera trapping rare and threatened avifauna in west-central Sumatra. *Bird Conservation International* 18(01): 30–37.

Fa, J.E., D. Currie, and J. Meeuwig. (2003). Bushmeat and food security in the Congo Basin: linkages between wildlife and people's future. *Environmental Conservation* 30(01): 71–78.

Fa, J.E., C. Peres, and J. Meeuwig. (2002). Bushmeat Exploitation in Tropical Forests: an Intercontinental Comparison. *Conservation Biology* 16(1): 232–237.

Hill, K., G. McMillan, and R. Fariña. (2003). Hunting-Related Changes in Game Encounter Rates from 1994 to 2001 in the Mbaracayu Reserve, Paraguay. *Conservation Biology* 17(5): 1312–1323.

Kelly, M. J. (2008). Design, evaluate, refine: camera trap studies for elusive species. *Animal Conservation* 11(3): 182–184.

Leader-Williams, N., S. Albon and P. Berry. (1990). Illegal exploitation of black rhinoceros and elephant populations: Patterns of decline, law enforcement and patrol effort in Luangwa Valley, Zambia. *Journal of Applied Ecology* 27(3): 1055–1087.

Lok, C. B., L. K. Shing, Z. Jian-Feng, and S. Wen-Ba. (2005). Notable bird records from Bawangling National Nature Reserve, Hainan Island, China. *Forktail* 21: 33.

Magurran, A. E. 2004. Measuring Biological Diversity. Blackwell Publishing.

Milner-Gulland, E.J., M.V. Kholodova, A. Bekenov, O.M. Bukreeva, A. Grachev, L. Amgalan and A.A. Lushchekina. (2001). Dramatic declines in saiga antelope populations. *Oryx* 35(4).

Milner-Gulland, E. J. and H. R. Akcakaya (2001). Sustainability indices for exploited populations. *Trends in Ecology and Evolution* 16(12): 686–692.

Milner-Gulland, E.J., E.L. Bennett and the SCB 2002 Annual Meeting Wild Meat Group. (2003). Wild meat: the bigger picture. *Trends in Ecology and Evolution* 18(7): 351–357.

Noss, A.J., R. Pena, and D.I. Rumiz. (2004). Camera trapping *Priodontes maximus* in the dry forests of Santa Cruz, Bolivia. *Endangered Species Update* 21: 43–52.

Noss, A.J., R. L. Cuéllar, J. Barrientos, L. Maffei, E. Cuéllar, R. Arispe, D. Rúmiz & K. Rivero. (2003). A camera trapping and radio telemetry study of lowland tapir (*Tapirus terrestris*) in Bolivian dry forests. *Tapir Conservation* 12(1): 24–32.

Novaro, A. J. (2004). Implications of the Spatial Structure of Game Populations for the Sustainability of Hunting in the Neotropics. People in Nature: Wildlife Conservation in South and Central America. K. Silvius, R. Bodmer and J. Fragoso, eds. New York, Columbia University Press.

O'Brien, T. G. (2008). On the use of automated cameras to estimate species richness for large-and medium-sized rainforest mammals. *Animal Conservation* 11(3): 179–181.

O'Brien, T.G., M.F. Kinnaird, and H.T. Wibisono. 2003. Crouching tigers, hidden prey: Sumatran tiger and prey populations in a tropical forest landscape. *Animal Conservation* 6: 131–139.

Payan, E. (2009). Hunting sustainability, species richness and carnivore conservation in Colombian Amazonia. PhD thesis, Department of Biology and Anthropology. London, University College London.

Peres, C. (1990). Effects of hunting on western Amazonian primate communities. *Biological Conservation* 54(1): 47–59.

Purvis, A., P.M. Agapow, J.L. Gittleman and G.M. Mace. (2000). Nonrandom Extinction and the Loss of Evolutionary History. *Science* 288(5464): 328.

Redford, K. H. (1992). The empty forest. *Bioscience* 42(6): 412–422.

Robinson, J. and E. Bennett (2004). Having your wildlife and eating it too: an analysis of hunting sustainability across tropical ecosystems. *Animal Conservation* 7(04): 397–408.

Robinson, J. G. and E. L. Bennett (2000). Hunting for Sustainability in Tropical Forests. New York, Columbia University Press.

Rovero, F., T. Jones, and J. Sanderson. (2005). Notes on Abbott's duiker (Cephalophus spadix True 1890) and other forest antelopes of Mwanihana Forest, Udzungwa Mountains, Tanzania, as revealed by camera-trapping and direct observations. *Tropical Zoology* 18(1): 13.

Rowcliffe, J. M. and C. Carbone (2008). Surveys using camera traps: are we looking to a brighter future? *Animal Conservation* 11(3): 185–186.

Silva, J. L. and S. D. Strahl. (1991). Human impact on populations of chachalacas, guans, and curassows (Galliformes: Cracidae) in Venezuela. Neotropical Wildlife Use and Conservation. J. G. Robinson and K. H. Redford. Chicago, Chicago University Press: 37–52.

Silvius, K.M., R.E. Bodmer, and J.M. Fragoso. (2004). People in Nature. Columbia University Press.

Strahl, S. and A. Grajal (1991). Conservation of large avian frugivores and the management of Neotropical protected areas. *Oryx* 25(1): 50–55.

Thiollay, J. (1989). Area Requirements for the Conservation of Rain Forest Raptors and Game Birds in French Guiana. *Conservation Biology* 3(2): 128–137.

Thiollay, J. M. (2005). Effects of hunting on guianan forest game birds. *Biodiversity and Conservation* 14(5): 1121–1135.

Tobler, M.W., S.E. Carrillo-Percastegui, R. Leite-Pitman, R. Mares, and G. Powell. (2008). An evaluation of camera traps for inventorying large-and medium-sized terrestrial rainforest mammals. *Animal Conservation* 11(0): 169–178.

Turvey, S. and C. Risley (2006). Modelling the extinction of Steller's sea cow. *Biology Letters* 2(1): 94–97.

Vickers, W. T. (1991). Hunting yields and game composition over ten years in an Amazon Indian territory. Neotropical Wildlife Use and Conservation. Chicago, University of Chicago Press: 53–81.

Wilkie, D.S., B. Curran, R. Tshombe and G.A. Morelli. (1998). Modeling the Sustainability of Subsistence Farming and Hunting in the Ituri Forest of Zaire. *Conservation Biology* 12(1): 137–147.

Yoccoz, N.G., J.D. Nichols and T. Boulinier. (2001). Monitoring of biological diversity in space and time. *Trends in Ecology and Evolution* 16(8): 446–453.

Yost, J. and P. Kelley (1983). Shotguns, blowguns, and spears: the analysis of technological efficiency. Adaptive Responses of Native Amazonians. R. Hames and W. T. Vickers. New York, Academic Press: 189–224.

Appendix. List of large mammals observed during the Kwamalasamutu RAP survey. CT = camera trap; K = Kutari; S = Sipaliwini; W = Werehpai. * Birds; included here for documentation of Trio names.

Scientific name	Common name	Trio name	Detection method	Site
Cuniculus paca	Paca	Kurimau	CT	K, S, W
Alouatta macconnelli	Guianan Red Howler Monkey	Aluatá	Heard	K
Ateles paniscus	Guianan Black Spider Monkey	Arimi; Tanonkonpe	Observed	K, S, W
Cebus apella	Brown Capuchin	Tarípi	Observed	K, S, W
Cebus olivaceus	Wedge-capped Capuchin	Ako	Observed	K, S, W
Chiropotes chiropotes	Guianan Bearded Saki	Isoimá	Observed	K
Dasyprocta leporina	Red-rumped Agouti	Akuri	CT	K, S, W
Dasypus kappleri	Great Long-nosed Armadillo	Kapai	CT	S
Dasypus novemcinctus	Nine-banded Armadillo	Kapai	CT	K, W
Eira barbara	Tayra	Ëkërëpukë	CT, observed	W
Leopardus pardalis	Ocelot	Pakoronko	CT	K, W
Mazama americana	Red Brocket Deer	Wikapao	CT, tracks	K, S, W
Mazama gouazoubira	Grey Brocket Deer	Kajaké	CT	K, S, W
Myoprocta acouchy	Red Acouchy	Pasinore	CT	K, S, W
Myrmecophaga tridactyla	Giant Anteater	Masiwë	CT	K
Nasua nasua	South American Coati	Seu	CT, observed	K, S, W
Panthera onca	Jaguar	Kaikui; Aturae	CT	W
Philander opposum	Common Gray Four-eyed Opossum	Aware	CT	K, S, W
Pithecia pithecia pithecia	White-faced Saki	Ariki	Observed	K
Priodontes maximus	Giant Armadillo	Morainmë	CT	S, W
Proechymis sp.	Spiny Rat	Kurimau	CT	K, S, W
Pteronura brasiliensis	Giant River Otter	Jawi	Observed	S
Puma concolor	Puma	Arawatanpa	CT, tracks	W, K
Saguinus midas	Golden-handed Tamarin	Makui	Observed	K, W
Saimiri sciureus sciureus	Guianan Squirrel Monkey	Karima; Akarima	Observed	K, S, W
Tapirus terrestris	Brazilian Tapir	Pai	CT, tracks	K, S, W
Tayassu pecari	White-lipped Peccary	Poneke	CT	W
Pecari tajacu	Collared Peccary	Pakira	CT, tracks, observed	K, S, W
Neacomys spp(?)	Mouse spp.		CT	S
Crax alector	Black Curassow*	Ohko	CT, observed	K, S, W
Crypturellus variegatus	Variegated Tinamou*	Sororsoroí	CT	K, W
Tinamus major	Great Tinamou*	Suwi	CT	K, W
Penelope spp.	Guan*	Marai	Observed	K
Psophia crepitans	Grey-winged Trumpeter*	Mami	CT	K, S, W